ROYAL NEW ZEALAND AIR FORCE

PER ARDUA AD ASTRA

Nmk
August 6, 1987

PORTRAIT OF AN AIR FORCE

The Royal New Zealand Air Force 1937–1987

ROYAL NEW ZEALAND AIR FORCE
PER ARDUA AD ASTRA

PORTRAIT OF AN AIR FORCE

The Royal New Zealand Air Force 1937–1987

Geoffrey Bentley
Maurice Conly

Grantham House
New Zealand

Also by Geoffrey Bentley
RNZAF A Short History 1969

Also by Maurice Conly
Ice on My Palette with Neville Peat 1977

Roundels of the RNZAF through the decades.

First Published 1987

GRANTHAM HOUSE PUBLISHING
P.O. Box 17-256
Wellington 5
New Zealand

ISBN 1 86934 010 8

Edited by Anna Rogers
Typeset by Wordset Enterprises Limited, Wellington
Designed by Graham Stewart of Bookprint Consultants
Printed in Hong Kong through Bookprint Consultants Limited, Wellington

Contents

Skyhawk pilot preparing for take-off.

Acknowledgements

Much of the material used in this book has been taken from departmental files and official documents, but the narrative has been filled out by the memories of former servicemen and women. To all who helped in this way, our grateful thanks.

The early chapters owe much to the original researches of Squadron Leader J. M. S. Ross and Wing Commander H. L. Thompson. The entire work has been given wider scope and greater authority by the considerable input of Wing Commander C. F. L. Jenks and Squadron Leader P. A. Harrison, acting as an editorial advisory committee.

Within the Air Staff at Defence Headquarters in Wellington, Group Captain P. J. Hughan, Director of Organisation and Staff Duties, and his directorate have been unfailingly helpful, in particular Squadron Leader M. J. House, liaison officer for the project, Flight Sergeant P. L. Bolton and Flight Sergeant G. Cole. The Defence Public Relations Directorate, first under Wing Commander G. T. Clarke and then under his successor Commander G. J. Power, gave the fullest co-operation.

The author gratefully acknowledges the considerable research assistance provided by Mrs Amanda Otzen, of the Defence Library staff in Wellington, and the typing and secretarial help from his wife. The artist acknowledgements the advice and assistance he received from his wife.

The Central Photographic Establishment at Ohakea, under Squadron Leader B. R. Anderson, the Defence Headquarters photographic section, and the base photographic sections at Auckland, Woodbourne and Wigram contributed most of the black and white illustrations. (Unless otherwise stated, the photographs reproduced in this book are from RNZAF official sources.) We are also deeply indebted to the many people throughout New Zealand who made historical photographs available from their own or family collections. Many of those from private albums could not be traced to their original sources.

Mr David Balm, of the Lands and Survey Department, Wellington, was most helpful in the production of maps, especially drawn for this book.

Geoffrey Bentley
Maurice Conly

Author's Note

Of the more than 55,000 New Zealanders who joined the Royal New Zealand Air Force during the Second World War, over 10,000 were transferred to the Northern Hemisphere where they served with the Royal Air Force in every theatre of war and in almost every unit. Their story has been fully told by Wing Commander H. L. Thompson in his three-volume official history *New Zealanders with the Royal Air Force*. Of the other 45,000, nearly 15,000 served at some time in the Pacific, while the remainder manned home establishments and trained those destined for overseas service. Squadron Leader J. M. S. Ross's official war history, *Royal New Zealand Air Force*, published in 1955, documents their story well.

All the events described in those volumes occurred before the RNZAF as a separate Service was nine years old. It is now 50 years of age and in the intervening decades has served its country — and others — in war and peace. It is this overall contribution that I have endeavoured to describe in *Portrait of an Air Force*, which marks the Golden Jubilee of the Service. The limitations of a single volume have meant that I have had to present the story of half a century in very condensed form, but the account is as accurate and authentic as I could make it.

The enormous contribution to the Royal Air Force by such distinguished New Zealanders as Marshal of the Royal Air Force the Lord Elworthy, Air Chief Marshal Sir Keith Park and Air Marshals Sir Roderick Carr, Sir Andrew McKee, Sir Hector McGregor and Sir Rochford Hughes, to mention but a few, have no part in this narrative because this is the story of New Zealand's own Air Force. I have touched briefly on RNZAF participation in the New Zealand squadrons of the RAF in the Second World War and have dealt a little more fully with the heroic stand against the Japanese by No. 488 Squadron in Singapore and the East Indies because this squadron was made up almost entirely of pilots and ground staff of the RNZAF.

I hope the reader will see behind every flying unit mentioned a much larger force of aircraft technicians, armourers, instrument repairers, suppliers, doctors, nurses, parachute packers, meteorologists, clerks, motor transport drivers, cooks, stewards and others who are all vital cogs in the Air Force machine. Not every serviceman in war emerges as a hero, for heroism requires opportunity and opportunity does not knock on all doors. But there was often heroic endeavour by RNZAF ground staff who gave long hours of selfless devotion to their given tasks, behind the scenes and beyond the spotlights of publicity.

Geoffrey Bentley

Skyhawks in an aerobatic manoeuvre called a 'bomb burst'.

Prologue

More than 120,000 people in 30,000 motor vehicles contributed to the worst traffic jam in New Zealand's history on Saturday 29 March 1958. Approaching the Manawatu area from Wellington in the south, Taranaki in the west, Hawke's Bay in the east and Waikato and the Bay of Plenty in the north, their common target was the air force station of Ohakea where the Royal New Zealand Air Force was about to celebrate its 21st birthday. Advance press and radio publicity promised the most spectacular air display ever seen in New Zealand.

As early as 8 a.m. there were indications in the Manawatu of the acute overcrowding on the roads that in a few hours would reduce all traffic to a mere crawling pace. Some motorists who had left Wellington at 7 a.m. for a journey that would normally take about two hours were still en route eight hours later. Others did not arrive at all but used their car radios to follow the proceedings. By 11 a.m. cars were stretched bumper to bumper along more than 100 kilometres of main highway. The frustrated motorists accepted the delay stoically; the prospect of a major flying display made the present inconvenience bearable.

The pageant mounted by the RNZAF that afternoon with the support of the Royal Air Force, the Royal Australian Air Force and the United States Air Force, besides the New Zealand Army's Special Air Service Squadron, held the record crowd enthralled. Super Sabre F-100 jet fighters of the USAF gave dramatic displays of supersonic flying; a Valiant bomber of the RAF flown by New Zealander Wing Commander L. H. Trent, VC, and USAF jet bombers and a Boeing KC-135 fuel tanker gave awe-inspiring performances; and the massed parachute drop by members of the SAS Squadron to a barrage of fire from 25-pounders and Bofors guns in a simulated aerodrome defence riveted the attention of the onlookers. But it was the superlative flying by the pilots of the RNZAF which won the most admiration from the 120,000 visitors to the station. The *New Zealand Herald* commented on the following Monday morning: 'Even the visiting American servicemen whistled with respect as the local pilots flew their outdated, outworked machines with inspired brilliance.'

This, it must be remembered, was 1958. In that year television had not arrived in New Zealand to satiate living room audiences with every conceivable kind of human endeavour, both real and fanciful, and the public went out in search of its entertainment. Nevertheless, the quality of the flying was high by any world standards, and the handling displays by the British V-bomber and a Sunderland flying-boat of the RNZAF, the latter only a few metres from the ground, were still being discussed years afterwards.

As a spectacular climax to the programme, four New Zealand Vampire pilots gave an exhilarating display of aerobatics in tight formation, weaving intricate patterns of smoke across the sky and concluding with a breathtaking 'thread the needle' act which was to earn plaudits for the team and its successors at subsequent pageants, such as that held to mark the opening of Wellington's new civil airport the following year. General Curtis E. LeMay, vice chief of the United States Air Force and a former head of its Strategic Air Command, told news media representatives later: 'That Vampire team ... put on the finest display of aerobatics I have seen in my life.'

Curtis LeMay was one of several high-ranking guests who watched the display from the VIP enclosure. Next to His Excellency the Viscount Cobham, GCMG, TD, Governor-General of New Zealand, sat Air Vice-Marshal C. E. Kay, CB, CBE, DFC, Chief of the Air Staff, RNZAF, and Mrs Kay, and with them were Air Marshal the Earl of Bandon, KBE, CB, CVO, DSO, RAF, Commander-in-Chief of Britain's Far East Air Force, and Countess Bandon. Also present was Australia's Chief of the Air Staff, Air Marshal F. R. W. Scherger, CB, CBE, DSO, AFC, who had flown in the previous day from Australia and presented the RNZAF with an anniversary gift from his own Service of a silver punch bowl and set of matching goblets. The Earl of Bandon had a special interest in the flying as his own Far East command included two RNZAF units — No.14 (Fighter) Squadron and No. 41 (Transport) Squadron — which were playing a part in anti-terrorist operations during the Malayan Emergency.

But there was one guest in the VIP gallery who had the best reason for being there and the greatest personal interest in seeing the Royal New Zealand Air Force at close quarters. This specially invited retired Royal Air Force officer, sitting with his prewar colleague, Air Vice-Marshal Sir Leonard Isitt, KBE, was Air Chief Marshal the Hon. Sir Ralph Cochrane, GBE, KCB, AFC. Britain might know him as the brilliant leader who, as Air Officer Commanding No. 5 Group, Bomber Command, directed 617 Squadron's famous 'dam buster' raids over Germany. To New Zealanders he was the man who laid the foundations for the Royal New Zealand Air Force. As Wing Commander the Hon. R. A. Cochrane, he was the officer sent by the Air Ministry in 1936 to investigate and report on this country's air forces, and it was on his recommendation that the Royal New Zealand Air Force was established as a separate branch of the armed forces with effect from 1 April 1937. From that date, Cochrane became the first Chief of Air Staff, in the rank of group captain.

He mused on this during the flying display at Ohakea that day in 1958, 21 years on, and later he had a close look at the excellent hangars and other buildings that he personally had proposed should be built for the young Air Force. During the next week he discussed those early days with senior officers at the Air Department. The two years he was in New Zealand were only a short period in his own long and distinguished career as an officer in the Royal Air Force, but it was a period for which the RNZAF has never ceased to be grateful.

Wing Commander the Hon. Ralph Cochrane was the right man at the right time for the emerging Royal New Zealand Air Force in 1936. Spare in build, with a slightly aloof manner and a keen intelligence, he had an incisive mind that was matched by a 'brook no nonsense' determination to get things done. He had a deep reverence for efficiency and he sought it uncompromisingly, even ruthlessly, according to some observers who have chronicled his wartime achievements such as his inspired leadership of No. 5 Group, RAF Bomber Command. 'Still probably the best brain in the Royal Air Force,' said British author Paul Brickhill a few years after the Second World War, when Cochrane was an air chief marshal, much decorated and ranking second in command of the vast Royal Air Force. It was a brain which this remarkable man, climbing rapidly to the top through his own ability, was to exercise in the best interests of New Zealand as the RNZAF's first Chief of Air Staff between 1937 and 1939.

Ralph Cochrane was born at Cults in Scotland on 24 February 1895, the youngest son of the first Lord Cochrane of Cults. He followed a strong family tradition by beginning his career in the Royal Navy but the air attracted him from the start. He was concerned with both airships and conventional aircraft during the First World War and graduated from Cranwell as an airship pilot in 1917. He was appointed to a permanent commission in the Royal Air Force in 1919 and was awarded the Air Force Cross the same year. His first tour of duty in the Middle East saw him Mentioned in Dispatches for operations in Kurdistan. His second tour there in 1928-29 gave him his first squadron command. Necessary elements to a well-balanced career — administrative duties at Cranwell, a course at the RAF Staff College and air staff duties at Headquarters of No. 7 Group — all followed and he had just completed a course at the Imperial Defence College when the call came which brought him to New Zealand in 1936.

To understand the state of New Zealand's air forces that year — the resources available to Cochrane in terms of military aircraft and trained men to fly and maintain them — it is necessary to look back more than 50 years, in fact to the year 1909 and to another remarkable man, one Henry Wigram.

Avro 626 NZ203 being restored by RNZAF technicians at Ohakea. This aircraft is now part of the Historic Flight of the Central Flying School based at Wigram.

Britannia over One Tree Hill, Auckland.

New Zealand's first military aircraft, the Bleriot XI *Britannia*, was given to the New Zealand Government in 1913 by the Imperial Air Fleet Committee in London, intended by the donors to be the nucleus of a New Zealand squadron. The aircraft was designed by Louis Bleriot for Gustav Hamel's historic 1913 flight from Dover to Cologne; he covered the 340 mile distance in four hours 18 minutes — a record at that time. The Bleriot's first New Zealand flight, on 17 January 1914, was made from Potter's Paddock at the foot of One Tree Hill, Auckland. After only three further flights the machine was dismantled and returned to storage in Wellington. When war broke out later that year the Bleriot was still awaiting its incorporation as an active unit of the New Zealand Defence Forces. Although Major General Alexander Godley, General Officer Commanding, recommended that the aircraft accompany the Expeditionary Forces overseas, it was dispatched instead to the Royal Flying Corps at Brooklands.

CHAPTER ONE

The Years of Indifference

There are good reasons why the RNZAF base in Christchurch, long known as 'the cradle of New Zealand military aviation', bears the name of Wigram. They are the same reasons why a large painting of Sir Henry Wigram by Archibald Nicoll has had a place of honour in the officers' mess at the base, serving to remind several generations of young officers of the crusading fervour of one man, whose foresight, tenacity and generosity paved the way for the formation of an air force.

But for Wigram's advocacy, New Zealand would have taken much longer to establish an air force and might never have acquired the aerodrome which became the birthplace of the New Zealand Permanent Air Force, later to be the RNZAF, and which is today still the centre of its initial flying training. No man contributed more to the birth of military aviation in this country in physical and mental effort, in inspiration and in gifts of land and money. Before he came on the scene, a few enthusiasts had made the first endeavours to build and fly aircraft in New Zealand — and their efforts were important and must not be discounted — but it was Sir Henry, then the Hon. H. F. Wigram, a member of the Legislative Council, who was the true 'father of New Zealand military aviation'. He was the first man with clear-cut views on the subject to advocate a definite air policy.

Born in London on 18 January 1857, Sir Henry Wigram was educated at Harrow and came out to New Zealand as a young man. He lost no time in establishing himself firmly in commerce. His civic works in Christchurch included a term as mayor, he served as a member of the Legislative Council for 17 years from 1903, he founded the Canterbury Seed Company and was a director of the *Lyttelton Times* and many other companies, he was the first president of the Canterbury Automobile Association and of the Canterbury Centre of the Royal Life Saving Society. Henry Francis Wigram was a visionary, a shrewd man of business and an outstanding administrator.

In 1909, addressing the Legislative Council, Wigram

Sir Henry Wigram, father of New Zealand military aviation. His portrait in oils by Archibald Nicoll has a place of honour at the Air Force base that bears his name.

The Sockburn airfield in the days of the Canterbury (NZ) Aviation Company, January 1918. Later it was to become RNZAF Base Wigram. *L. Hinge*

urged the formation of a flying corps as part of the nation's defence forces. His colleagues showed little interest in the proposal. Eight years were to pass before, with the stimulus of war with Germany, his campaigning led to the establishment of the Canterbury (NZ) Aviation Company, one of the country's first flying schools. Wigram was well equipped to spearhead such a campaign. His standing in local and national government, his business acumen, his newspaper interests and his financial resources were all valuable weapons in the personal armoury with which he fought his crusade against official apathy and inertia. In one objective of his mission — the establishment of the Canterbury (NZ) Aviation Company — he had the active support of a group of Christchurch businessmen. He died on 6 May 1934, after a full life, but not before making generous gifts of both land and money to establish the Sockburn airfield which in 1923 was renamed Wigram. The Air Force has never forgotten him. Lady Wigram, who lived on to a great age, was always remembered when special occasions came around.

'There is a growing feeling that the time has arrived when some special effort should be made to strengthen the resources of the Empire in aerial craft in view of the great advances being made in this direction by foreign powers.' These words were part of a letter addressed to the New Zealand High Commissioner in London early in 1913 by a group of patriotic British businessmen. They had offered to give the latest type of military aeroplane to one of the principal countries of the British Empire as the start of a great Imperial air fleet. The New Zealand Government accepted gratefully. Although some years were to pass before anything positive was done about a New Zealand air force, the words of that 1913 letter heralded the beginning of military aviation in this country.

Subsequently, a two-seat Bleriot monoplane, appropriately named *Britannia*, powered by an 80 hp Gnome engine and capable of a speed of 70 mph in still air, was shipped to New Zealand, arriving at Wellington in the *Athenic* on 29 September 1913. Its arrival posed some problems. First, there was no one in the defence forces competent to fly or service it. A letter had to be sent to the Aviation School at Farnborough in England, inquiring how the aircraft could best be used; 'I presume a shed will be necessary?' wrote the Minister of Defence, Mr J. Allen.

Britannia was put on show and made several demonstration flights at Auckland with a civilian pilot, Mr J. J. Hammond, of Feilding, at the controls. The first flight was on 17 January 1914. The last was on 28 January that year, when the first woman to be carried in a heavier-than-air machine in New Zealand flew as a passenger. This had not been authorised, however, and Hammond (who had returned to New Zealand from Australia to fly the Bleriot) was relieved of this duty. *Britannia* was offered to the British Army Council on 23 September 1914. The offer was accepted and the aircraft was shipped to the United Kingdom on 13 October.

With the coming of the First World War, Wigram's voice was heard again in the Legislative Chamber. The Government, he said, must establish flying schools to train pilots for the Royal Flying Corps and to prepare New Zealand for the formation of its own aviation corps in the interests of national defence. Concerned more with the demand for infantry, the Government rejected his arguments, and the formation of flying schools was left to private enterprise. In this way, New Zealand military aviation got away to an uncertain start.

Two schools of flying training emerged: one, the New Zealand Flying School, at Kohimarama, Auckland, and the other, formed by the Canterbury (NZ) Aviation Company Ltd, at Sockburn, near Christchurch.

The New Zealand Flying School was launched by two Auckland motor engineers, Leo and Vivian Walsh, in October 1915. They were resourceful men in an age when people created their own pastimes. Although they had been well educated at King's College, and had immense enthusiasm, they had little knowledge to aid their ambition. Lacking both finance and aviation experience, they had four years earlier built an aircraft and taught themselves to fly it. (The first time that Vivian saw a plane in flight was when his first pupil went solo!) Determination and patience, combined with a natural aptitude, took them a long way along a road which other men in other parts of the world were travelling to reach the same goal.

The Walsh brothers first attempted to build a land plane but after some setbacks they tried their hands instead at a seaplane of the Curtiss type which first flew on 1 January 1915. The machine was a success and became the nucleus of their school, which they expanded with other machines built by their own hands. Spurred on by the prospect of success, they introduced flying instructors and aero mechanics, training these men themselves. They were rare men, these Walshes, and true pioneers. They enjoyed no State grants or subsidies — Leo approached the Government for assistance and was offered 'moral support' — and they had to depend on public-spirited citizens for finance. One of these backers, Mr R. A. Dexter, of Auckland, became a director of the company.

The Walsh brothers (and, later, the Canterbury Aviation Company) gave to New Zealand the distinction of being the only Commonwealth country where pilots were trained for and by private enterprise. Undaunted by their own government's apathy, the brothers had made an offer to the British Government to train pilots for the Royal Flying Corps, and received the reply that all suitable candidates qualifying in New Zealand for the Royal Aero Club's certificate would be accepted for RFC commissions. First-class fares to the United Kingdom would be met and a refund of £75 towards the cost of training would be made to each candidate on arrival in Britain. With aircraft emerging as a potent force in warfare, Britain's Royal Flying Corps was eager for as many trained men as this country could send. The British Government was unequivocal: 'Send as many pilots as possible,' it said in effect.

By the end of the First World War 110 pilots had been trained at the New Zealand Flying School. Of them, 68 were commissioned in the Royal Flying Corps (or the Royal Air Force) and eight of them were to win the Distinguished Flying Cross. The school had another reason for pride. Not one fatal flying accident occurred to mar its training record.

A Curtiss flying-boat B over the New Zealand Flying School, Kohimarama.

George Bolt.

Opposite page: C. M. Hill, chief instructor at the Canterbury Aviation Company's flying school at Sockburn, in the cockpit of a Caudron.

When ill health forced Vivian Walsh to abandon active flying in 1918, a young man named George Bolt, who had learned to fly at the school, was appointed chief instructor. At the age of 18 George Bolt had built a glider with the help of a friend, Merville Angus, and they made several successful flights in Christchurch. A biplane which carried the pilot suspended between the two wings, it was claimed to be the first glider of its type in New Zealand.

The flying training school at Sockburn, operated by the Canterbury (NZ) Aviation Co. Ltd, was formed in 1916 and flying began there the following year. Its principals had sought in vain to persuade the New Zealand Government that financial aid was in the national interest and so, like the Walsh brothers, they entered into an agreement with the British Government. Training commenced on 19 June 1917, with one aircraft, a Caudron two-seater biplane with a 60 hp Anzani engine. C. M. Hill, who had come from the Hall Flying School at Hendon, England, was the instructor and the ground staff included J. G. Mackie and J. E. Moore, responsible for airframes and engines respectively. When Hill took the company's chairman, Henry Wigram, aloft in the Caudron that same day, it was the first passenger flight in the history of the aerodrome. By the end of the war there were two more instructors on the staff, B. Dawson, on loan from the Walsh school at Kohimarama, and J. C. Mercer, who had been taught to fly by Hill.

Altogether, 170 pilots were trained at the Sockburn school by the end of the First World War. Of these, 156 went overseas and were commissioned in the Royal Flying Corps, Royal Naval Air Service or Royal Air Force. From their ranks and those of the Walsh brothers' school came many of the first members of the New Zealand air force.

Besides the men who trained at Kohimarama and Sockburn, countless other New Zealanders served in the First World War in the Royal Naval Air Service, or the Royal Flying Corps or the Royal Air Force (which was created when the RNAS and RFC were amalgamated on 1 April 1918) either as pilots or ground staff. Many of them retained their interest in flying and were later attracted to the New Zealand air force in its infancy.

A Caudron biplane of the Canterbury Aviation Company with pupil.

Flying-boats of the New Zealand Flying School at Kohimarama in 1919.

An Avro 504K of the New Zealand Permanent Air Force over Sockburn in 1923.

A flight of Bristol Fighters, early 1930s, with Wigram below.

CHAPTER TWO

The Influence of Bettington and Salmond

Because of the years of indifference, the First World War came to an end without any form of military air organisation existing in New Zealand. Among the politicians, however, there was some stirring of conscience or awakening of interest, or both (fostered, no doubt, by the numbers of combat pilots returning home), for the British Government was asked to make available an officer who could advise on New Zealand aviation policy. The importance of the aeroplane as a weapon of war had been well illustrated in the previous four years, and the return of trained pilots presented a perfect opportunity for the formation of at least the nucleus of an air force.

The British Government responded promptly to the request and on 19 March 1919 Colonel A. V. Bettington, DSO, RAF, arrived in New Zealand in the *Athenic* with two RAF mechanics and four aircraft (two Bristol F2B Fighters and two DH4 bombers), his staff being augmented by three New Zealand officers who had served in the RAF during the war. They were Major A. de Bathe Brandon,* DSO, MC, RAF, a Wellington barrister, Captain J. H. Don, a Timaru dental surgeon, and Lieutenant E. T. Shand, a Dunedin sheep farmer. The aircraft were based on the Canterbury Aviation Company's airfield at Sockburn where the Government had erected two hangars. The first flight was undertaken by Captain Don in one of the Bristols on 4 September 1919.

The report which Colonel Bettington submitted to the New Zealand Government on 5 June 1919, after making a thorough assessment of the country's air defence needs, showed remarkable foresight. He had no faith in the likelihood of an enduring peace and prophesied that the Pacific would become a trouble centre, pointing out that Japan had emerged from the war immeasurably stronger than before and with a growing population. New Zealand, he contended, should develop closer defence ties with Australia. He was not led astray, as were some later planners, by the existence of the supposedly 'impregnable' base at Singapore.

Bettington recommended that the Government should at once form an air arm of trained men returning from RAF service and take over Sockburn as a training centre. He maintained that the minimum force required for an efficient operational air force should comprise two flying-boat squadrons, two fighter squadrons, two bomber squadrons, one torpedo-bomber squadron, an aircraft depot and two aircraft parks. All these units should be formed immediately in cadre, capable of rapid expansion. In four years, he argued, the air force should have a permanent establishment of 369 officers and airmen. The total cost during the first four years was to be £1,294,000. He also proposed a Territorial Air Force of 1134 men to be created within eight years.

* Brandon was already a man with a reputation. In 1916, as a second lieutenant, he had assisted in the shooting down of two raiding Zeppelins over England.

Colonel A. V. Bettington. *Canterbury Public Library*

Air Marshal Sir John Salmond. *From a painting by Orde*

Major T. M. Wilkes (left) and Captain L. M. Isitt in front of a DH4 at Timaru in 1920. *Talma Studio, Timaru*

While Bettington was formulating his recommendations, the New Zealand Government received an offer from Britain of a gift of 100 aircraft to help in the establishment of an air force. The types offered included Avro 504Ks, Bristol Fighters, Dolphins, DH9s, DH9As, Salamanders and SE5s.

In the political climate still prevailing Colonel Bettington's scheme was considered too ambitious and he was asked to submit a modified plan. He did so in August that year but this was also unacceptable and he returned to Great Britain feeling that he had not been able to accomplish much in New Zealand.

The aircraft which Bettington had brought with him were retained at Sockburn and in 1919 a young Christchurch officer just back from RAF service, Captain L. M. Isitt, was posted to Sockburn to replace Captain Don. It was Isitt's duty to care for the aircraft, act as liaison officer between the Canterbury Aviation Company and the Government and to supervise military flying training if and when it took place.

Following Bettington's return to England, the Government set up its own advisory committee to make recommendations on aviation generally and to consider acceptance of Britain's offer of aircraft. It was decided at length to accept 20 Avro 504Ks and nine DH9As and to request six flying-boats instead of the balance of the 100 aircraft originally offered. The request was dispatched to Britain in January 1920. Two weeks later came the reply from the Air Ministry: owing to the long delay between its offer being made (11 July 1919) and its acceptance (28 January 1920), no surplus DH9As remained, nor could any flying-boats be supplied.

The New Zealand Government took prompt steps to reserve what aircraft were still available and 29 machines were later shipped to the Dominion — 20 Avro 504Ks and nine DH9s — which, together with the two Bristol F2B Fighters and the two DH4s left behind by Bettington, gave a fleet of 33. Only six machines were retained by the Government, however, and these were stationed at

An Avro 504K takes off on a refresher course training flight.

The first airman to win the Victoria Cross was London-born Second Lieutenant William Barnard Rhodes-Moorhouse, whose grandfather was an early colonial settler and whose grandmother was of Maori blood. Rhodes-Moorhouse, who was educated at Harrow and at Trinity College, Cambridge, qualified as a pilot in 1911 and three years later joined the Royal Flying Corps. In May 1915 he took part in a major bombing raid at Courtrai in which he came down to 300 feet to find his target, and was hit by rifle and machine-gun fire. Although mortally wounded he flew back to base and insisted on making his report before being admitted to hospital where he died 24 hours later. His commanding officer wrote: 'His last act has set a standard of courage which others may hope to equal but never excel.' The VC was awarded posthumously.

Sockburn for military purposes. The remainder (18 Avros and all of the DH9s) were lent to various civilian flying enterprises being developed in many parts of the country.

By 1920 there was general agreement that, in view of the probable development of flying, the future defence pattern for New Zealand would embrace the use of air, as well as land and sea forces. That year a Defence Committee was formed to ensure the effective co-ordination of the three arms, but as the first meeting, held in July, was also the last, perhaps co-ordination was not then regarded as so important. An Air Board was, however, established 'to advise the Minister of Defence on all matters of aviation'.

The Air Board included representatives of such Government agencies as might conceivably have an interest in military aviation — Defence, Post and Telegraph, Public Works and Lands and Survey Departments. Its members were: Major General Sir E. W. Chaytor, General Officer Commanding New Zealand Forces (president), Brigadier General G. S. Richardson, Commander T. A. Williams, RN (naval adviser), Messrs T. N. Brodrick (Lands and Survey), A. Markman and G. McNamara (Post and Telegraph Department) and F. W. Furkert (Public Works Department). When the composition of the board was promulgated, the press observed caustically that not one of its members knew anything about flying! To rectify this, a qualified pilot, Captain T. M. Wilkes,* New Zealand Staff Corps, was appointed secretary. He was an experienced combat pilot with service in the RFC and the RAF, the only officer of the New Zealand Staff Corps so qualified.

What may be regarded as New Zealand's first positive aviation policy received the seal of government approval on 25 September 1920. It stated (in part):

> The Government (is) to make provision for the development of Aviation along lines which will enable the Dominion to possess civil aviation for commercial and other needs and at the same time provide for the necessities of aerial defence in case of emergency.
>
> The (Air) Board (is) to make recommendations as to contracts to be entered into for the carriage of mails, passengers, etc. All contracts (are) to be submitted for approval to the Minister in charge and the Postmaster-General and to be confirmed by both.

* Later Group Captain T. M. Wilkes, CBE, MC.

A Hucks starter, mounted on a surplus RAF vehicle, is used to start a DH9 of the NZPAF.

Attention (is) to be paid to meteorological conditions and the Board is to recommend in what direction assistance should be given to the Meteorological Department with a view to equipment to meet the needs of aviation.

While policy was being formulated in Wellington the fate of the two flying schools at Auckland and Christchurch was hanging in the balance. The coming of peace had deprived them of their regular income and, with no more military pilots to train, their future was uncertain. Colonel Bettington had suggested that the Canterbury Aviation Company should undertake initial training of pilots and mechanics for the New Zealand Air Force. He had not considered the school at Kohimarama necessary for air force training and had proposed that the Government should recognise its excellent work during the war by buying its equipment.

Bettington's recommendations were not acted on immediately. Instead, the schools were subsidised by grants and both were employed on experimental airmail services. From February 1923, refresher training was given by the Canterbury (NZ) Aviation Company to former RAF pilots, under Captain Isitt's supervision.

The refresher courses held at Wigram were characteristically easygoing affairs and the pilots, all of whom had flown in the First World War, attended in civilian dress and brought with them their old wartime flying kit. The indomitable spirit of these men was the slender thread which held the embryonic air force together during this difficult period. The refresher courses may not have done much actual good, but they kept alive the spirit of military aviation in New Zealand.

A refresher course for officers was also provided at Kohimarama in 1923. This was the only military use made of the school and it closed down the following year. The Government purchased its equipment, but made little use of it. In any case many of the aircraft were worn out.

To the air force in the making, fell the task of controlling civil aviation besides supervising service training and maintaining military aircraft. Duties included the inspection of civil aircraft and aerodromes and the granting of licences to pilots.

The year 1923 saw more positive progress towards the establishment of an air arm: first, the New Zealand Aviation Corps was created as part of the Army and then, with effect from 14 June, the New Zealand Permanent Air Force (NZPAF) was formed as a part of the New Zealand Military Forces and the New Zealand Air Force (NZAF) was constituted as part of the territorial forces. Both were nevertheless controlled by the Army. In retrospect, the two air forces and their similarity of titles must have caused confusion in the minds of the general public.

At the time of its formation, the NZPAF comprised two officers and three airmen, with 15 aircraft (seven Avro 504Ks, four DH9s, two DH4s, and two Bristol Fighters). To accommodate the new air force, a military aerodrome was required. The New Zealand Government, influenced and encouraged by the Hon. Henry Wigram, purchased the assets of the Canterbury Aviation Company (land, buildings, aircraft and spares, horses and implements), an act facilitated by Wigram's gift of £10,000, which was in fact a reduction of the purchase price of the land. Appropriately, the airfield was renamed Wigram Aerodrome.

By the end of 1923 the strength of the NZPAF was four officers and five other ranks. Major Wilkes was in command at Defence Headquarters in Wellington and Captain Isitt commanded Wigram. To assist him, Captain J. L. Findlay, MC, was appointed as flying instructor and Lieutenant T. J. Denton was transferred from the Army to be full-time equipment officer. Three other ranks were enlisted or transferred for maintenance and equipment duties.

DH4s and a DH9 at Wigram in the early 1920s.

Meanwhile, the territorial NZAF consisted of 72 officers,* who had served as pilots in the First World War. Many, notably Keith Caldwell, Maurice Buckley, H. B. Burrell, Gerry Stedman, John Seabrook, Sidney Wallingford, 'Tiny' White and George Carter, were to play important roles and later hold high office in the RNZAF which they helped to develop. Some, like Ian Keith, became prominent flying instructors. Two of them, Lieutenants Hood and Moncrieff, were to lose their lives in the historic first attempt to fly the Tasman Sea.

In 1925 an area of 167 acres was purchased at Hobsonville, near Auckland, and here was established in 1929 the seaplane base and airfield which have since served the RNZAF so well. Captain Isitt was told that he would be given command of the new Hobsonville station when it was completed, so, in 1926, command of Wigram was handed over to Captain Findlay and Isitt went to England, to spend two years there gaining experience in flying-boat handling, reconnaissance, organisation and administration.

* Major Keith L. Caldwell, MC, DFC; Captains R. E. Buckingham, MC, C. F. Meagher, AFC, M. W. Buckley, H. B. Burrell, C. C. L. Dowdall, J. L. Findlay, MC, G. L. Stedman, H. F. S. Drewitt, MC, DFC, P. K. Fowler, N. E. Chandler, John Seabrook, AFC, A. F. Upham, DFC, M. C. McGregor, DFC, F. W. Crawford; Lieutenants R. A. Stedman, F. J. Horrell, John Coates, C. H. Noble-Campbell, AFC, George Hood, C. A. Umbers, K. W. J. Hall, G. V. T. Thomson, H. R. T. Hughes, Sidney Wallingford, Ivo Carr, F. de M. Hyde, AFC, W. F. Parke, S. T. Goodman, MC, DCM, T. W. White, G. G. A. Martin, A. C. McCarthur, A. H. Skinner, G. J. Wilde, AFC, K. J. Gould, D. C. Inglis, DCM, I. L. Knight, W. R. Patey, DFC, T. B. Hardy, R. J. Thompson, F. S. Gordon, DFC, H. C. Lloyd; Second Lieutenants M. D. Sinclair, P. Mitford-Burgess, G. Cotton-Stapleton, J. E. Stevens, I. E. Rawnsley, George Carter, H. I. N. Melville, Frank Hazlett, J. R. Richardson, Dennis Winfield, D. G. Gregorie, N. F. Harston, I. A. McGregor, R. B. Reynolds, W. B. Gillespie, H. N. Hawker, J. W. H. Lett, E. A. F. Wilding, W. E. Norton, M. H. Otway, W. G. Coull, R. G. MacDonald, R. J. Sinclair, R. J. M. Webber, W. W. Withell, R. C. Hancock, I. H. N. Keith, J. R. Moncrieff, E. D. Williams. Captain Christopher Musgrave, AFC, RAF Reserve (Class C) was attached.

HRH the Duke of York (later King George VI) inspects a refresher course at Wigram for NZAF pilots early in 1927, in the course of the Royal Visit. From left to right: Major T. M. Wilkes, Director of Air Services, Army Office; the ADC to the Duke of York; Captain J. L. Findlay; the Duke; K. Hall, W. F. Parke, George Bolt, Harry Lett, J. R. Moncrieff (shaking hands with the Duke), C. Craig and I. E. (Jack) Rawnsley. (Lieutenant Moncrieff was lost at sea on 10 January 1928 with Captain G. Hood, in the first New Zealand attempt to fly the Tasman Sea.)

In 1928 Air Marshal Sir John Salmond, a senior Royal Air Force officer (who, as Marshal of the RAF Sir John Salmond, GCB, CMG, CVO, DSO, would become Britain's Chief of Air Staff two years later), visited New Zealand to advise on air defence. At that time the NZPAF consisted of five officers and 17 airmen, two stations (one still under construction) and 14 aircraft. More Bristol Fighters had been acquired, two in 1925 and three in 1927, and another six Avro 504Ks had been purchased, arriving in the second half of 1925. A de Havilland DH50A had been acquired early in 1927, for aerial survey work, and three Gloster Grebes* had been ordered — Sir Henry Wigram had assisted the purchase with a gift of £2,500 — and the last two aircraft came to hand in September when Sir John Salmond arrived in the country. Many of the earlier aircraft had by then been withdrawn from service.

Sir John Salmond recommended that the NZPAF should be built up to a strength of nine permanent units,+ with an immediate increase in establishment to 218 officers and airmen. He also recommended a scheme for short-service commissions in both the NZPAF and the RAF. The Depression of the 1930s prevented Sir John's major proposals from taking effect but some others bore fruit.

In February 1929 the Navy and Army held combined exercises in the Auckland area and Captain Isitt flew a series of observation and photographic missions from the partly completed aerodrome at Hobsonville. The Minister of Defence, the Hon. T. Wilford, was a passenger on many of these flights, the first recorded of a joint land-air-sea exercise in New Zealand. (In addition to refresher training, the NZPAF had, during the years 1924-30, co-operated with both the Army and the Navy in their major exercises, and air surveys had been flown for local authorities and for the State Hydro Department.)

Towards the end of 1929 Captain Wilkes was transferred to London to serve as New Zealand's liaison officer at the Air Ministry. (Isitt had acted in this capacity during his two years in Britain but Wilkes's was the first full-time appointment in this post.) Wing Commander Stuart Grant-Dalton, DSO and Bar, AFC, who had recently resigned from the RAF, applied for the post of Director of Air Services and was duly appointed to fill the vacancy. Not a great deal is known of Grant-Dalton today but a territorial pilot who flew with the NZAF between 1929 and 1932, Flight Lieutenant John Seabrook, was a great admirer of the man and he wrote in subsequent years: 'Greater recognition should be made of the immeasurable contribution to the progress of the air force by Wing Commander Stuart Grant-Dalton as Director of Air Services. "G. D." inspired loyalty and enthusiasm and he was a great natural leader in those early years.' Grant-Dalton introduced RAF ranks and methods into the NZPAF in December 1929 in lieu of Army titles and practices. Although Army uniforms continued to be worn until 1931, air force blue was issued for use 'on special occasions'.

The expansion recommended by Sir John Salmond was now being reflected in new aircraft arrivals. By the end of 1929 five of seven de Havilland DH60 Moth aircraft had arrived; the remaining two would follow in 1930. They were used as elementary trainers and two were fitted for operating as seaplanes. Three Fairey IIIF aircraft were ordered for general reconnaissance and Navy co-operation and two arrived in September 1929. A Saunders Roe Cutty Sark amphibian was ordered and arrived in May 1930. The flying-boat was used for various tasks including Army and Navy co-operation, aerial photography and air transport.

Early in 1930 the NZPAF enlisted 15 aircraft apprentices who, after eight weeks' general service training at the Army's Trentham camp, were posted to Wigram. This had the effect of almost doubling the strength of the PAF.

In 1930 the territorial New Zealand Air Force was organised into a wing of four squadrons, with squadron headquarters in each of the four main centres, Auckland, Wellington, Christchurch and Dunedin. Because the wing, commanded by Wing Commander K. L. Caldwell, MC, DFC, had an officer establishment only, the aircraft and ground staff were made available by the NZPAF.

In January 1930 the New Zealand Permanent Air Force carried out its first active operations. In response to a local uprising in Western Samoa, New Zealand dispatched HMS *Dunedin* in an endeavour to restore order, and on to the ship went a DH60 Moth seaplane with Flight Lieutenant S. Wallingford as pilot and two air force corporals to service the machine. (Wallingford, who was 31 years of age, had joined the PAF the previous year, after concluding a short-service commission in the RAF.) The expedition spent two months in Samoa during which time Wallingford flew 90 hours in the Moth, in general reconnaissance, message dropping and ground co-operation flights. Because this was active service, the Moth had fittings for a Lewis gun mounted on each side of the cockpit. The aircraft was flown with one hand and the gun aimed and fired with the other. To make up for weaponry deficiencies in the *Dunedin*'s 'air arm', the ship's torpedo officer set about improvising a bomb for young Wallingford. He put 2 lb of gun cotton in a treacle tin; the cork of a whisky bottle separated the striker from the percussion cap.

One day Flight Lieutenant Wallingford decided that, as his bullets were having no effect on what he considered were rebel boats, he would drop the 'bomb'. He pulled out the cork and dropped the tin over the side. It did not explode, but there was a formal complaint later — from the local missionary boat!

This was the first aircraft dispatched from New Zealand on 'warlike operations'. But there were no newsreel or television cameramen in those days to record the historic events for audiences at home, and Wallingford's part in the Mau uprising of 1930 was watched, if at all, only by his fellow servicemen.

After the Murchison earthquake of 1929 and again

* The Gloster Grebe was the first single-seat fighter in service in New Zealand, although one was a two-seat version.

\+ At Christchurch, a flying training school, a fighter flight, an Army co-operation squadron and a stores section; at Auckland, a coastal reconnaissance flight, a torpedo-bomber flight, a fighter flight and a stores section; at Blenheim, a torpedo-bomber flight.

A Fairey IIIF on the ramp at Hobsonville.

Wing Commander S. Grant-Dalton, DSO, AFC.

The NZPAF's first active operation, Western Samoa, January 1930.

Hawker Tomtit trainers over Wigram in 1931.

following the disastrous Napier earthquake in 1931, air force pilots, together with aero club members, did valuable work in maintaining communications and flying medical and other supplies into the affected areas.

In February 1931 the Permanent Air Force took delivery of four Hawker Tomtit aircraft. These machines were used as advanced trainers for territorial pilots who had been taught to fly in Moths, before allowing them to fly the larger Grebes and Bristol Fighters. In August that year a de Havilland DH80A was added to the inventory. Known as the Puss Moth, this aircraft replaced the DH50A for aerial survey work and it was also used as a light utility aircraft and for VIP transport. (The Duke of Gloucester who, with the Duchess of Gloucester, visited New Zealand in 1934-35, was carried in the Puss Moth on at least one occasion.)

There were significant events in the mid 1930s in many parts of the world. In 1933 Germany withdrew first from the Disarmament Conference and then from the League of Nations. In 1934 Britain was forced to see her disarmament policy as dangerously idealistic and her Prime Minister, Stanley Baldwin, announced that the strength of the RAF was to be increased by 41 squadrons in five years! In 1935 Britain's Secretary of State for Air, Lord Londonderry, disclosed that the programme was being accelerated and would provide for 49 new squadrons.

The middle of the decade also saw the end of the Depression and this, coupled with the march of events in Europe, resulted in more attention being focused on the strength of the air force in New Zealand.

The Annual Report to Parliament by Major General Sir William Sinclair-Burgess, KBE, CB, CMG, DSO, NZSC, General Officer Commanding the New Zealand Army, for the year 1934-35, was a clear indication of expansion for the NZPAF which henceforth would be known as the Royal New Zealand Air Force, King George V having approved this new designation which was gazetted in February 1934. (The RNZAF was still a component of the New Zealand Army, however.) The annual report said the air vote for the year, £197,934, was a considerable increase compared with that of the previous year but unfortunately it had been underspent by £65,704. This serious situation was brought about by the fact that building work was not begun until the latter half of the financial year 'and also because of the impossibility of getting aircraft deliveries at short notice'. Of the unexpended balance, buildings accounted for £32,410 and aircraft £28,845.

'The strength of the RNZAF at 31 May 1935 was 10 officers and 101 other ranks, one additional officer and 57 airmen having been enlisted during the year,' said the report. Flight Lieutenant A. de T. Nevill, RNZAF, had left for England in June 1934 to attend the RAF Staff College at Andover, and to be the RNZAF's liaison officer at the Air Ministry. The territorial wing had a strength of 86 officers plus four New Zealand Medical Corps officers and one artillery officer, 15 new appointments having been made during the year.

In March 1935 the RNZAF took delivery of the first of 12 Vickers Vildebeest II bombers which had been ordered the previous year. They were a good choice for the RNZAF at that time, and although there were criticisms that they were not the fastest or most modern aircraft of their type, they were destined to play a major part in New Zealand's war effort at home in the early years of the war that was to come. The bombers were apportioned evenly between Wigram and Hobsonville where bomber reconnaissance flights were established. These units were the first properly constituted operational elements of the RNZAF.

Moth seaplanes and Fairey IIIFs from Hobsonville co-operated with the Navy in torpedo, gunnery and anti-aircraft exercises as a regular requirement until 1936 when *Achilles* and *Leander* joined the New Zealand Division of the Royal Navy. These ships carried their own Walrus aircraft and the Hobsonville machines were less in demand. But co-operation continued in joint exercises with both the Army and the Navy. By mid 1936 Hobsonville had two aircraft hangars, administration offices, stores, engine and airframe repair shops, a garage, new single airmen's barracks and a number of married men's quarters. A concrete slipway had also been built there.

There was expansion at Wigram, too. The buildings taken over from the Canterbury Aviation Company were no longer adequate and two new hangars were being erected. The airfield had been extended, a development made possible by Sir Henry Wigram's further gift, in December 1931, of 81 acres of adjoining land, formerly occupied by the Canterbury Park Trotting Club. Married men's quarters were also being provided.

The year 1936 also saw the acquisition of four Avro 626 trainers, which had been ordered the previous year. A distinct advance on existing trainers, the 626 was one of the first of the multi-role aircraft capable of a variety of training tasks from piloting to fixed and free gunnery, bomb aiming and stretcher bearing. It had been developed from the Tutor, the RAF's standard trainer of the early 1930s. The Avro 626s entered service with A Flight of the Flying Training School at Wigram where they were used for intermediate training before pilots advanced on to the Vildebeests of B Flight. Late in 1936 the Bristol Fighters were phased out of their flying training role and relegated to ground instructional use.

The RNZAF was now emerging from the doldrums into which government apathy had permitted it to drift for too many years. The new Labour Government, which had been elected in 1935 under Prime Minister M. J. Savage, was more alive to the needs of military aviation than had been its predecessors and in the light of quickening events

The first of 12 Vickers Vildebeest bombers purchased by the RNZAF, at Weybridge, England, awaiting a demonstration flight on 31 January 1935 before Sir James Parr, New Zealand High Commissioner in the United Kingdom. Eighth from the left in the group of onlookers is Flight Lieutenant A. de T. Nevill, RNZAF Liaison Officer at the Air Ministry, London. *London News Agency Photos Limited*

Avro 626 advanced trainers over Quail Island, Banks Peninsula, circa 1936.

An early refresher course for territorial pilots at Wigram in the 1930s. The aircraft is a Hawker Tomtit, an elementary and intermediate trainer. Pilots identified are, from left, standing: C. M. Duthie, A. C. Parsons, M. W. Buckley, G. L. Stedman, Gibbons, Weston, unknown, unknown, I. E. Rawnsley and M. K. Archibald. Seated, at far right, is W. F. Parke.

in Europe the term 'air power' was acquiring a new relevance.

Once more the British Government was asked to make available an officer capable of studying New Zealand's air power resources and assessing the country's needs for an effective air force. Eight years before Britain had sent an air marshal, a man destined to be Chief of the Air Staff in his own country. Those members of the RNZAF who remembered Sir John Salmond's visit might have felt somewhat let down when they learned that a wing commander — the air force equivalent of a lieutenant colonel — had been briefed for the task. They need not have worried. They would learn soon enough that dynamic leadership was at hand and that an exciting three years of expansion and progress lay before them.

CHAPTER THREE

Cochrane the Architect

Early in 1936 the New Zealand Government decided to establish the Royal New Zealand Air Force as an independent service divorced from Army control, and Squadron Leader T. M. Wilkes, who held the appointment of Director of Air Services, was asked to prepare a scheme and assess the costs. Although he had long been a staunch advocate of an independent air force, Wilkes refused to accept the task, arguing that any worthwhile scheme would be costly and that if he or any other New Zealander drew it up it would almost certainly be rejected on the grounds of finance. If the Government really wanted a separate air arm, he argued, then let it follow a familiar precedent and obtain the services of an expert from Great Britain. The recommendations that such a person would make would have unquestionable authority and the positive development of an air force — without too much quibbling over finance — was more likely to follow, he said.

The Hon. Sir Ralph Cochrane, a portrait taken when he was an Air Chief Marshal in the Royal Air Force. *Air Ministry*

In this advocacy Wilkes showed great astuteness, and those who acknowledge a debt to Cochrane, the architect of today's RNZAF, can also be grateful to Wilkes who, by his stubborn stand, was indirectly responsible for Cochrane's appointment. At a press conference in New Zealand during his 1958 visit, Air Chief Marshal the Hon. Sir Ralph Cochrane paid tribute to Wilkes for his pioneering endeavours. 'I don't think Group Captain Wilkes has ever been given the credit he deserved,' he told newsmen. 'His farsightedness and plain hard work in those early days made a real contribution to the formation of the RNZAF.'

By 1936 it was obvious that war with Germany was at least a danger to be reckoned with. It was equally obvious that the aeroplane was destined to play a major part in Empire defence. Confronted with the New Zealand Government's request for an RAF officer to advise on the country's air defence needs, Britain's Air Ministry selected Wing Commander the Hon. R. A. Cochrane, AFC, who had just completed a course at the Imperial Defence College. He was a very good choice, as subsequent events were to show.

On the sea voyage to New Zealand, Cochrane began his evaluation, based on the briefing he had received, together with his own study of New Zealand's geographical position. He arrived in Wellington on 4 November 1936,

Vickers Vildebeest Mk III bombers.

eager to begin his mission. After paying the usual formal calls he had lengthy talks with senior officers and then travelled around the country surveying the resources available. It did not take long to assess them. There were the two bases at Wigram and Hobsonville with a total staff of just over 100 men. There were 12 Vickers Vildebeest bombers and an assortment of aircraft including two Gloster Grebe fighters, one DH60 Gipsy Moth trainer, four Hawker Tomtit trainers, four Avro 626* trainers, one DH80 Puss Moth communications and survey aircraft and two Fairey IIIF bombers.

Taking the view that New Zealand's defence involved (a) local defence, (b) the defence of shipping routes, and (c) the security of the United Kingdom, Cochrane suggested that the RNZAF should be capable of countering raids by cruisers, armed merchantmen or submarines, and by aircraft carried in such ships. New Zealand was not likely to be in danger of invasion as long as Singapore was maintained as a major base and the British fleet could be sent to the Pacific theatre in the event of war. If these two assumptions proved invalid, no forces New Zealand could afford to maintain would be strong enough to deal with a major attack. Raiding forces would probably need to secure bases in the Pacific Islands and New Zealand should be prepared to protect both potential Pacific bases and vital communication points in the area from enemy attack.

Considering all these factors, Cochrane recommended that the RNZAF should maintain two medium bomber squadrons capable of locating and attacking enemy raiders before they reached New Zealand's coast. These squadrons should be equipped with aircraft which could range to bases in the South Pacific or to the RAF's Singapore bases if necessary. His proposals were:

1. The RNZAF should be constituted as a separate service controlled by an Air Board under the Minister of Defence.
2. An establishment of two permanent squadrons equipped with medium bombers, the total first-line strength to be 24 aircraft, together with repair facilities and reserves of aircraft.
3. Reserves of personnel to be instituted on a basis of numbers required to maintain two medium bomber squadrons and one Army co-operation squadron in a condition of readiness for a major war. They should be trained to full operational standard.
4. Civil aviation to be encouraged and aero clubs supported.
5. The United Kingdom to be invited to co-operate in developing Pacific bases.

He proposed that the question of Territorial squadrons should be reconsidered later.

To put these recommendations into effect, Cochrane suggested that all aircraft and personnel be concentrated at Wigram to make an immediate start in training the men required for the Air Force and the Air Force Reserve. Extra accommodation should be provided as quickly as possible and additional training aircraft obtained. The next step, between 1937 and 1939, should be the construction of accommodation, repair facilities and bomb storage for two medium bomber squadrons. Finally, in 1938-39, equipment and reserves for the two squadrons should be purchased.

He estimated that the capital cost of this scheme would be £1,124,000. The value of aerodromes and buildings already in existence could be deducted from this sum but the nett cost would still be £1,100,000. This included the building of a permanent station for the two bomber squadrons, the cost of the aircraft and reserves, the provision of bombs and bomb storage, additional construction at Wigram, the provision of landing grounds in the Pacific Islands and equipment for wireless telegraphic communications. The estimated annual cost of the scheme would be £435,000.

Cochrane's recommendations resulted in the passing of the Air Force Act of 1937. With effect from 1 April that year, the RNZAF became a separate branch of the Dominion's defence forces. The Air Department Act passed at the same time set up a State department which would administer both civil and military aviation.

There was widespread approval within the RNZAF and elsewhere when Cochrane was invited to remain in New Zealand and implement his own programme. He pondered the invitation. He thought of Wilkes and Isitt, the two most senior officers in the RNZAF and both obviously in line for advancement to key positions. He consulted them both. On receiving their assurances that they had no objections to a newcomer stepping in 'over their heads', he accepted the offer. The Air Ministry cabled its concurrence and his appointment as Chief of Air Staff with the rank of group captain was effective from 1 April 1937.

The new Air Department followed closely the lines of Britain's Air Ministry and, as could be expected, most aspects of the RNZAF adhered fairly strictly to the RAF pattern for many years to come.

The Chairman of the Air Board was the Minister of Defence, the Hon. F. Jones. Cochrane as Chief of the Air Staff was supported at the Air Department by Wing Commander Wilkes as Chief Staff Officer (Equipment)

The Air Department building, Wellington, in 1937.

* A further four Avro 626s had been ordered before Cochrane's arrival but arrangements fell through and they never arrived.

Gloster Grebe NZ502
of No. 1 FTS,
Wigram, circa 1937.

Mr T. A. Barrow, CBE, JP.
Appointed the first Air Secretary on 1 April 1937, he occupied the post until 1953.

and by Squadron Leader Isitt as Chief Staff Officer (Personnel). Mr T. A. Barrow was appointed the first Air Secretary. Wing Commander Wilkes was also appointed Controller of Civil Aviation (Civil Aviation Branch of the Air Department) with Flying Officer J. M. Buckeridge as his senior assistant and Mr R. C. Kean, AMI Mech.E, as Chief Inspector of Aircraft. These appointments were all effective from 1 April 1937. At this stage Wilkes held his two appointments concurrently.

On 1 September Wing Commander Wilkes relinquished the appointment of Chief Staff Officer (Equipment) while retaining the appointment of Controller of Civil Aviation and on 1 November he was promoted to group captain. On 1 December Isitt, newly promoted to wing commander (the promotion effective from 1 April), was appointed to the newly created post of Air Force Member for Personnel, and Flight Lieutenant A. de T. Nevill, newly promoted to squadron leader with effect from 1 April, was appointed Air Force Member for Supply.

The formation of a separate branch of the Air Department to administer civil aviation was a logical step as was the appointment of Wilkes as Controller of Civil Aviation. In that capacity he sat as a member of the Air Board on all discussions which involved both civil and

RNZAF Station Wigram in 1938. A major expansion programme was to follow.

Prewar hangar construction at Ohakea.

Officers of the Wellington Squadron, Territorial Air Force, in early 1938. Front row (left to right): Flight Lieutenant I. E. Rawnsley, Flying Officer J. Te H. Grace, Squadron Leader E. A. Gibson (Officer Commanding), Flying Officer G. Carter (Engineer Officer) and Flying Officer L. F. P. Taylor. Middle row: Flying Officer R. B. Morphy, Pilot Officer J. G. Hogan, Pilot Officer A. M. Prichard, Captain J. M. Twhigg, NZMC, Flying Officer G. Martin, Pilot Officer R. A. W. Coker, Flying Officer S. R. Meatchem. Back row: Flying Officer H. C. Walker, Flying Officer Wilton, Flying Officer M. K. Archibald, Pilot Officer E. C. Smart, Flying Officer R. F. Smillie. *S. C. Smith*

military aviation. This organisation enabled a pooling of such resources as were required by both branches of the department — meteorological services, wireless and navigation aids, aerodromes and emergency landing fields.

So that there would be a core of experienced officers to help build up the new RNZAF, several RAF officers were lent to New Zealand during 1937–39 and, in exchange, RNZAF officers were attached to the RAF to gain experience. Details of the exchange scheme had been worked out in 1926 during the Imperial Conference, when Squadron Leader Isitt had met Sir Philip Game, then Air Force Member for Personnel in the RAF. More than 10 years later the existence of a plan made it easier to carry through the exchanges. A number of New Zealanders serving with the RAF were selected for specialist courses in signals, navigation and armament, and then transferred to the RNZAF and returned to New Zealand.

Early in 1937 a start was made in putting Cochrane's recommendations into effect. Wigram was reorganised into a flying training school with an annual output of 40 pilots, later to be increased to 80, and the training of pilots for the RNZAF and RAF began. Hobsonville was converted to a stores and repair base and also became a training school for ground personnel. The station was enlarged by the purchase of an additional 55 acres of land.

After a survey of possible aerodrome sites for the bomber squadrons, it was eventually decided to build two stations instead of one and land was bought at Whenuapai, near Hobsonville, and Ohakea, near Palmerston North. Orders were placed in Britain for 30 Vickers Wellington twin-engined bombers and supplies of ammunition and bombs. The approved programme envisaged a regular air force of 100 officers and 900 airmen, compared with the total complement of just over 100 men that Cochrane found on his arrival. A Civil Reserve of Pilots would be formed with the co-operation of the aero clubs and there would be the additional strength of the Territorial Air Force.

Five Airspeed Oxfords were ordered, four as trainers and one for aerial survey. These were the forerunners of a total of 299 aircraft of this type which would go into service in the years immediately ahead. In 1937, when Cochrane placed the order, the Oxford was going into service with the RAF as its first twin-engine, low-wing monoplane advanced trainer, and it was clearly a very suitable aeroplane on which to train crews who would fly the Wellington bombers when they arrived in New Zealand.

Some second-hand Vickers Vincent bombers were also ordered from the RAF — the only external difference between the Vincent and the Vildebeest was the length of the tailplane struts — besides 29 second-hand Blackburn Baffins (which had seen service in Royal Navy carriers) with which to equip the territorial squadrons.

Supplementary expansion programmes approved over

Vickers Vincent advanced trainers of No. 2 FTS leaving Woodbourne in formation, about 1940.

Working on this Baffin aircraft at Woodbourne in late 1939 are, from left: Aircraftmen Bill Danks (with plug spanner in hand), Todd, R. F. Ward, unknown, George Patton (in cockpit), Brown and Oxley. Danks, Todd, Ward and Patton were Permanent Staff; the others were territorials. *Gordon McCusker*

Airspeed Oxfords at Wigram.

the next two years provided for schools to train flight mechanics and flight riggers at the railway workshops in the four main centres, territorial squadrons in the same four cities and the establishment of a regular squadron at Blenheim. The two-station air force was growing. Wigram and Hobsonville were expanding visibly and the fine, big stations that were to blossom at Ohakea, Whenuapai and Woodbourne (Blenheim) were already in sight.

In 1939, his term ended, Cochrane handed over the reins to another RAF officer on loan and went back to the United Kingdom and a brilliant career as one of Britain's great leaders in the then imminent air war. Cochrane had much to show for his term as Chief of Air Staff. The permanent buildings which stand on the RNZAF's major bases were a fine monument, 50 years later, to his vision and drive. He saw to it that the foundations of a first-class fighting force were well and truly laid. He established a pattern for an air force that would have quality in all it did, in pure flying, aircraft maintenance and administration, and this quality was apparent in almost everything that the RNZAF was to accomplish in the years that followed.

Blackburn Baffins of the Wellington Territorial Squadron on a prewar patrol over the Marlborough Sounds.

Flying Officer F. R. Dix in Vildebeest NZ105 passes Mount Cook en route to Westland on an early survey mission, 1936.

Fairey Gordons were in service as advanced trainers from 1939 to 1941 when they were replaced by Harvards.

Tiger Moths being assembled at the de Havilland factory at Rongotai.

CHAPTER FOUR

Prewar Expansion under Saunders

The impetus given to the development of the RNZAF by Group Captain Cochrane was not lost in the changeover of command. His successor as Chief of Air Staff was Group Captain H. W. L. Saunders,* a South African by birth who, after Army service in his own country (then part of the British Empire), had joined the Royal Flying Corps in 1917 and made notable progress in the Royal Air Force as a career officer. Saunders was an extremely capable administrator and a natural leader of men, and in his new command he continued the pattern of distinguished leadership which Cochrane had set. He had been in office only a short time before a plan for a new and important expansion programme for the RNZAF was presented to the New Zealand Government and approved in April 1939.

Group Captain H. W. L. Saunders, MC, DFC, MM, the RNZAF's second Chief of the Air Staff.

This programme provided that the new air base being constructed at Blenheim should become a second flying training school with an output of 140 pilots a year and that the existing flying training school at Wigram be enlarged to achieve a similar output. It also required the purchase of more aircraft and equipment to facilitate the expansion of flying training at these two bases.

In the years immediately before the outbreak of the Second World War the development of flying training had two main objectives: to train pilots for the expanding RNZAF, and to provide pilots for service in the RAF. Since 1919 a number of New Zealanders had travelled to England at their own expense to join the RAF. Some had obtained short-service commissions while others had entered the service under a scheme whereby a limited number of candidates from the Dominions were accepted annually at the RAF Cadet College at Cranwell, to be appointed later to permanent commissions.

* Later Air Chief Marshal Sir Hugh Saunders, GCB, KBE, MC, DFC, MM, RAF.

In 1934, when the RAF was beginning to expand, the Air Ministry had suggested that New Zealand should train a number of pilots each year and send them for four years' service with the RAF after which they would return home and serve a further period on the Reserve. The New Zealand Government would be responsible for their training and their passage to England, and would be paid £1,550 for each pilot sent. New Zealand was unable to adopt this suggestion in 1934 because of the lack of training equipment that year but the scheme was put into operation in the middle of 1937. At the same time another scheme was started under which a number of candidates were selected each year and medically examined in New Zealand for short-service commissions in the RAF. The candidates were accepted by the RAF without further ado and their fares to England were paid by the Air Ministry.

By the time Group Captain Saunders was implementing his expanded flying training programme, a total of 133 officers had been trained at Wigram in just over two years and posted to the RAF for short-service commissions.*

In that same month of September, 1938, the New Zealand Government, in view of the deteriorating international situation, proposed if war should come to set up an organisation to train 1000 pilots a year for the RAF. The offer was accepted in principle by the British Government and plans were drawn up to provide for the necessary establishment. An important step before the outbreak of war was the agreement reached between the two Governments that a factory should be provided in New Zealand for the production of training aircraft. The de Havilland Aircraft Company of New Zealand Ltd was incorporated in March 1939, and in a factory built on the site of the present Wellington Airport production of Tiger Moths from completely knocked down kitsets began early in 1940.

* Most of this training had been directed by Squadron Leader E. G. Olson, chief flying instructor at Wigram, and Flight Lieutenants F. R. Newell and R. J. ('Nugget') Cohen, his two flight commanders. (Olson, Cohen and Newell had all come to the RNZAF after service in the RAF.) Olson was moved to Hobsonville in 1938 and Cohen became CFI at Wigram. Flight Lieutenant Newell was posted to England in September the same year for a course at the RAF Staff College and was replaced by Flight Lieutenant D. W. Baird who, following service with the RAF, had joined the RNZAF that year.

The RNZAF's first major air display was the Rongotai air pageant of 1938. Officers and men are about to be inspected by the Governor-General, Lord Galway (second from left). At left is Squadron Leader E. A. Gibson and third from left is Group Captain the Hon. R. A. Cochrane. *Evening Post*

The Pacific Defence Conference of 1939 declared Fiji a vital area in New Zealand's defence. This conference, held in Wellington in April and attended by representatives from the British, Australian and New Zealand Governments, expressed views similar to those already propounded by Cochrane. If British reverses in Europe were to prevent the British Fleet from coming east, it said in effect, Japan might conceivably be able to mount a large attack against Australia and New Zealand. It was considered, however, that the probable tactics of attack would be raids against shipping by cruisers, armed merchantmen and submarines; raids by armed merchant cruisers with landing parties for the temporary occupation of islands in the Pacific for refuelling bases or the destruction of cable stations; and cruiser raids against ports in Australia and New Zealand and perhaps some Pacific islands. Attacks would probably be limited to this scale only if Singapore could be made secure, and the conference considered that the governments concerned should be ready to meet attacks on a greater scale. To attempt an invasion of Australia or New Zealand, the Japanese would need to secure South Pacific bases, said the conference, and it recommended that New Zealand should take immediate steps to protect potential bases.

Certainly, the most vital strategic area in the South Pacific, from New Zealand's point of view, was Fiji. The harbour facilities of Suva and the oil fuel supplies held there made it one of the most important naval fuelling bases in the South-West Pacific. Moreover, it had a major cable and wireless station, and it was bound to become more and more important as a centre of air communications. The conference decided that New Zealand should immediately provide two landing fields on Viti Levu, the principal island of the group, one near Suva and the other on the north-western coast, and that part of New Zealand's reserves of fuel, bombs and ammunition should be stored there. It was agreed that Tonga should also be surveyed with a view to providing landing fields and that New Zealand should assume responsibility for this and for air reconnaissance over the area embraced by the New Hebrides, Fiji and Tonga.

As a result of the Pacific Defence Conference, New Zealand's plan to provide wartime training for 1000 pilots annually was amended. The alternative proposal, which would provide about 650 pilots, 300 observers (navigators) and 350 air gunners, was accepted by the Air Ministry in May 1939.

By mid 1939 there was much to occupy the minds of those men who controlled the destiny of the small but growing RNZAF. The 30 new Vickers Wellington bombers ordered as a result of Cochrane's two-bomber squadron concept were being readied in England for ferrying to New Zealand. The big Ohakea and Whenuapai stations were nearing completion and the new Woodbourne station was emerging. New schools were urgently required for recruit training and for the training of flying instructors, observers and air gunners. The important developments in the Fiji region were imminent. Life was full of interest for men of all ranks and there was a feeling of anticipation in the air.

The Territorial Air Force, too, was progressing at last. The Wellington squadron had been formed by Flight Lieutenant I. E. Rawnsley at Cochrane's request before the first Baffins were received in March 1938, and had undergone training at an inaugural camp at Omaka under Rawnsley's command. It was now exercising with New Zealand naval ships and had Squadron Leader Gibson as its commanding officer.

The Christchurch squadron, under Squadron Leader G. L. Stedman, was complete with its own permanent maintenance staff, had taken delivery of its first Baffin aircraft in September 1938 and had established a pattern of training camps. Squadron Leader D. M. Allan's Auckland squadron had received its Baffins early in 1939 and was now well advanced with training at Hobsonville in navigation, ship recognition, bombing and search and patrol procedures. The Dunedin squadron, although it had been authorised, had not then been formed.

In June and July, with war imminent, every available TAF pilot was sent to Wigram to undergo a two months' course of general reconnaissance. At the same time, the Civil Reserve of Pilots was expanded. This reserve, which had been instituted in 1937, was open to candidates who had attained the required standard of education and physical fitness and who were willing to serve in the RNZAF in emergency. The Government had agreed to pay for the initial flying training, which was fixed at 40 hours in the first year and two refresher courses each of 10 hours in the succeeding years. Training was being provided by the aero clubs which were also doing valuable work in training pilots for the RNZAF. The Civil Reserve scheme was co-ordinated by Flight Lieutenant H. B. Burrell under the general supervision of Group Captain Isitt.

The Civil Reserve of Ground Staff was also reexamined in late 1939. A register of tradesmen and potential administrative and technical officers who could be called on to serve in the RNZAF in emergency, the scheme had been set up in September 1938 and after the Munich crisis was accelerated. It was widely publicised through the news media and by posters and circulars sent to employers. Applications were called for air force trades such as wireless operator, electrician, instrument maker, coppersmith, sheetmetal worker, fitter gunsmith, radio mechanic, photographer, fabric worker, cabinet maker, clerk and storeman.

The expansion of the RNZAF was now moving at a rapid pace. In the South Island the erection of buildings had begun at Taieri, near Dunedin, where a major RNZAF training station would shortly emerge and where the fourth territorial squadron would be located.

On 1 September 1939 the Governor-General declared a State of Emergency and issued two proclamations of vital interest to the RNZAF; one transferred the Reserve to the Regular Air Force and the other made the Territorial Air Force liable for continuous service. The strength of the RNZAF at that stage, exclusive of Reserves, was: Regular Air Force: 91 officers, 665 airmen; Territorial Air Force: 79 officers, 325 airmen.

Tiger Moth elementary trainers flying in formation.

Group Captain K. L. Caldwell, MC, DFC.

Vildebeest and Vincent bombers being serviced in the early 1940s.

Tiger Moths, Harvards and Oxfords at the Central Flying School, Tauranga.

CHAPTER FIVE

A Call to Arms

At dawn on 1 September 1939, dive bombers of Hitler's Luftwaffe attacked Poland and several towns and cities, including the capital, Warsaw, were bombed. Simultaneously, the German Army crossed the Polish border and engaged the brave but pitifully weak defending forces.

That day Britain and France delivered an ultimatum to Germany calling on her to cease aggression against Poland and to prepare to withdraw her armed forces, otherwise they would fulfil their pledges to go to Poland's aid. On 3 September, no such undertaking having been received by the British Prime Minister, Mr Neville Chamberlain, Britain and France were officially at war with Germany. In New Zealand a Cabinet decision was made on the night of 3 September to go to war against Germany on the side of Britain and France.

One immediate result of the outbreak of war was that the RNZAF's 30 new Wellington bombers were never to be seen in New Zealand. A few days before the official announcement the New Zealand Government offered the bombers and their crews to the RAF as an immediate contribution to the war against Germany. The offer was accepted. (As a result, New Zealand was without modern operational aircraft until the arrival of the first Hudsons in 1941.)

On the day that war was declared, the RNZAF was ordered to mobilise. First, the three territorial squadrons were called up for service and brought up to strength.

The Auckland territorial squadron mobilised at Hobsonville under Flight Lieutenant C. L. Monckton, a Waipukurau sheep farmer, who acted as OC for a few days until Squadron Leader W. G. Coull, a Christchurch businessman, arrived to take command. Intensive training began at once and the squadron's Baffins were kept at instant readiness because of the possibility of enemy submarines entering the Hauraki Gulf.

The Christchurch squadron, also equipped with Baffins, reported for duty at Wigram on 4 September and stood by for operations from that day. Squadron Leader G. L. Stedman, the OC, was chief instructor at the Canterbury Aero Club and could not then be released for full-time active service so Flight Lieutenant G. N. Roberts* was given command.

Three days after war had been declared the Wellington squadron moved its Baffins across Cook Strait to Woodbourne to give air protection to a vital shipping lane. Since Squadron Leader Gibson was required for other duties, Squadron Leader R. J. Sinclair, a Pahiatua farmer, was given command. The squadron operated as a general reconnaissance unit and carried out submarine patrols and shipping escort duties over the approaches to Wellington Harbour.

* Later Air Commodore Sir Geoffrey Roberts, CBE, AFC.

Blackburn Baffin NZ157. Twenty-nine of these aircraft were purchased in 1937 to equip the four territorial squadrons. They had earlier seen service on Royal Navy aircraft carriers.

The Second World War was to embrace every continent save South America and every ocean; and members of the Royal New Zealand Air Force would be found fighting in every theatre, from Britain and Europe, North Africa and the Middle East to South-East Asia and the South Pacific. They fought in the air over the Atlantic, Indian and Pacific Oceans and the Mediterranean Sea. Nearer home, they made a substantial contribution to the Allied effort which later turned back the Japanese advance in the South Pacific.

To understand why so many members of the RNZAF, besides the 500 and more New Zealanders who had joined the RAF, were found in every theatre of war when the RNZAF's own operational commands were confined to the South-West Pacific, one must consider the Empire Air Training Scheme (EATS), to which New Zealand was a contributor. It was largely as a result of this gigantic and eminently successful undertaking that New Zealand airmen were scattered to the far corners of the globe.

The first few weeks of war convinced Britain that air supremacy was essential for her survival and that she must greatly expand her air forces. It was assumed that 50,000 aircrew would be needed each year to develop the RAF and clearly Britain alone could not train this number, so it was proposed that 50 flying schools should be established elsewhere in the Commonwealth. Canada, with her vast land area, her industrial capacity and her proximity to the resources of the United States, was an obvious choice for a large part of such a plan.

Late in September a scheme for Empire air training was put to the governments concerned and it was decided that air missions from Britain, Australia and New Zealand should visit Canada to discuss it. In the New Zealand delegation were the Chief of Air Staff, Group Captain Saunders, and the Air Secretary, Mr Barrow. The mission planned well. The task which New Zealand undertook was the provision annually of 880 fully trained pilots for the RAF besides 520 pilots, 546 observers and 936 air gunners trained to initial stages who would receive their advanced training in Canada, making an annual total of 2882 aircrew.

The principal objective of the Empire Air Training Scheme* (also known as the British Commonwealth Air Training Plan) was to establish a training organisation that would help create air forces superior in size and quality to those of the enemy. And as the schools turned out their graduates in tens of thousands so British air superiority increased in every theatre of war. EATS produced a great co-ordinated network of flying training extending from New Zealand-based stations such as Ashburton in mid Canterbury, via Canada to the United Kingdom and thence to every storm centre of the war in the air.

A total of 7002 New Zealand airmen would graduate from Canadian aircrew training, and in all, 131,553 airmen would be trained in Canada for flying duties under the EATS commitment. In Canada, the scheme ran from 29 April 1940 until 31 March 1945.

As part of the scheme, the British Government undertook that all aircrew from the Dominions would be identified with their respective countries. Altogether, seven squadrons with a New Zealand identity were subsequently formed within the RAF, six of them under EATS. Their aircrew members were largely New Zealanders and many of the squadrons were commanded by New Zealanders. In addition to these seven squadrons, there was New Zealand representation, and often command, in almost every RAF unit which served during the war.

In order to meet its commitments under the EATS, the RNZAF at home had to set up an organisation capable of accepting for training every four weeks 144 pilots for elementary flying training, 80 pilots for advanced training and 42 observers and 72 wireless operator/air gunners for initial training. Existing flying training schools had to be expanded and others provided.

Throughout 1939 and 1940 training went ahead despite the shortages of instructors, aircraft and equipment. About 76 light aircraft were taken over from the aero clubs and private owners and 50 were used for elementary training until Tiger Moths began to flow from the production lines at Rongotai or reach New Zealand by ship from the United Kingdom. At the same time, work proceeded on the construction of the new training schools. Many of the young men who went to Canada in the first drafts after partial training with the RNZAF were the products of its elementary flying training schools (EFTSs).

One of the first requirements, a flying instructors' school, was formed at the Auckland Aero Club's aerodrome at Mangere on 10 September 1939, under the command of Squadron Leader D. M. Allan, the aero club's chief instructor. In the preceding two months Wing Commander G. S. Hodson — an English officer on loan from the RAF — and Squadron Leader M. F. Calder — commanding officer and CFI at Wigram — had visited all the aero clubs in New Zealand and tested their flying instructors. As a result the instructors had been given refresher courses at the Auckland Aero Club to bring their procedures up to date so that the standard of the club's Civil Reserve and RNZAF training carried out by the clubs could be improved.

Squadron Leader Stedman had been given command of No. 1 EFTS at Taieri when it formed in October 1939, and he had the services of Flying Officer A. W. Burbidge as chief flying instructor. Training was given in de Havilland Tiger Moths taken over from civilian use. The maintenance staff consisted of Flying Officer W. Temple, Sergeant J. J. Simpson and 18 untrained airmen. In the first few days of the school's operations Temple and Simpson were fully extended and had very little sleep.

No. 2 EFTS, formed at New Plymouth in November 1939, had Flight Lieutenant A. C. Upham as its first commanding officer. Its initial equipment comprised former civilian aircraft types other than Tiger Moths and its first instructors were men who had served with aero clubs.

RNZAF Station Harewood opened at Christchurch in May 1940 under the command of Wing Commander Sir

* Group Captain T. W. White, CBE, ED, RNZAF, was Head of New Zealand's Air Mission in Ottawa and New Zealand's representative on the EATS Supervisory Board.

Robert Clark-Hall* and in August No. 3 EFTS accepted its first intake. By this time military Tiger Moths were available in numbers and were to serve all the EFTSs for the remainder of their existence. No. 4 EFTS was formed at Whenuapai in December 1940.

Advanced training for non-Canada bound pilots was completed initially at the Flying Training School at Wigram but another similar school, No. 2 FTS, opened at Woodbourne towards the end of 1939. The station commander was Wing Commander K. L. Caldwell†, who had graduated from the first course of pilots at the Walsh brothers' school at Kohimarama in 1915 and who had earned both the Military Cross and the DFC in the First World War. Chief flying instructor was Squadron Leader B. S. Nicholl. On 18 September RNZAF Station Ohakea was officially opened with Flight Lieutenant E. G. Gedge, MC, an Englishman, as its first commanding officer. Its main function at first was to train observers and air gunners but for the first few weeks of its existence it served also as a recruit training depot.

The very rapid expansion of the flying training organisation required a corresponding increase in the training of ground technicians. New technical training schools sprang up, the Central Trade Test Board was instituted to examine airmen at the end of their trade training and recruit training depots were further developed. RNZAF Stations Levin, Hobsonville and Rongotai — the last named utilising the temporary buildings erected for the 1940 Centennial Exhibition — accommodated much of this increased activity.

Early in 1940 trained technicians became available from courses organised in the country's four railway workshops. These courses had been introduced in July 1939, when training began in the Hutt workshops. Technical training centres were then opened in the Hillside (Dunedin), Addington (Christchurch) and Otahuhu (Auckland) railway workshops. The men were well trained and the Air Force soon felt the benefit. The scheme continued until September 1940 when the increased facilities for technical training within the RNZAF made the service more self-sufficient in this respect. The railway workshops training produced a total of 595 flight riggers and flight mechanics over approximately 14 months.

In January 1940, the Electrical and Wireless School was formally established at Wigram. The training of wireless operators and wireless mechanics had actually begun in January 1938 and when war began there were trained RNZAF wireless operators in England who had been sent there to fly out with the Wellington bombers. Training on a wartime basis had begun in October 1939, but as there was insufficient space at Wigram it was carried out at the School of Engineering at Canterbury University College, Christchurch, the trainees being accommodated at Rolleston House. The building of suitable premises at

At the top of the loop in a Tiger Moth.

* Air Marshal Sir Robert Clark-Hall, KBE, CMG, DSO. On his retirement from the RAF in 1934 he came to New Zealand. On the outbreak of war he volunteered his services to the RNZAF and accepted a temporary commission in the rank of wing commander. Later promoted to air commodore, he was AOC Southern Group 1943-44 and AOC No. 1 (Islands) Group 1944-45; retired 1946.

† Air Commodore K. L. Caldwell, CBE, MC, DFC and Bar.

One of the civil DH89s (Dragon Rapides) impressed into Air Force service after war was declared.

NZ151

Newly assembled Airspeed Oxfords at RNZAF Station Hobsonville.

Wigram facilitated the development of the E and W School from 1940 onward, covering all branches of communications, as part of No. 2 Technical Training Depot (later No. 2 Technical Training School), formed in September 1939 to train wireless operators, wireless electrical mechanics, instrument makers and repairers, armourers and fitter armourers. The E and W School was to play a major part in the service in the years ahead, its officers and instructors earning the widest respect.

Other units followed. An Administrative Training School was formed at Wigram to produce clerks and equipment assistants; and the Initial Training School was formed at Rongotai as a recruiting centre for pilots, observers and air gunners. This school was later transferred to Levin.

The essential expansion of the Air Force at this time was helped by an immediate response on the part of New Zealand men following the outbreak of war. As soon as it was made known that recruits were required there was a flood of applications for enrolment as pilots, observers and air gunners, and by the end of March 1940, seven months after the outbreak of war, the strength of the RNZAF had increased to 385 officers and 3600 airmen.

Supermarine Walrus amphibians first equipped the cruisers *Achilles* and *Leander* of the New Zealand Division of the Royal Navy in 1936 and 1937. They were withdrawn in 1942, and the Royal New Zealand Air Force took over some of them for air sea rescue duties and the training of flying-boat pilots. At the right, L2222 is seen being hoisted aboard one of the cruisers. This aircraft later served with the RNZAF as NZ151, as shown on left.

DH86s and DH89s of the Air Gunners and Air Observers School, Ohakea, about 1940.

Officers of the New Zealand Flight, RAF Station Marham, England, August 1939. From left: Flying Officer J. N. Collins, Flight Lieutenant C. C. Hunter, Flying Officer J. Adams, Pilot Officer T. O. Freeman, Squadron Leader C. E. Kay, Squadron Leader M. W. Buckley, Flying Officer A. B. Greenaway, Flying Officer A. A. N. Breckon, Flying Officer N. Williams, Flying Officer F. J. Lucas, Pilot Officer H. W. Coleman and Pilot Officer W. M. C. Williams.

Wellington IC AA-N at Feltwell in the spring of 1941, being 'bombed up' by No. 75 Squadron ground crew before a mission. *Air Ministry*

CHAPTER SIX

No. 75 Squadron in Europe

When, with war clearly imminent, the New Zealand Government waived its claim to the 30 new Wellington bombers and offered the aircraft and their crews to the RAF, no one could have foreseen that the result would be the formation of an élite combat unit which would not only earn an unparalleled war record in the Royal Air Force but also achieve a lasting place in the history of the Royal New Zealand Air Force.

The story of 75 Squadron is an integral part of the story of the RNZAF because it was formed with RNZAF crews, all handpicked for the new bomber force which had been part of Cochrane's concept for the defence of New Zealand. These crews, who had travelled to England specifically to fly the new Wellingtons home, became instead the nucleus of No. 75 Squadron. The Air Ministry gave the squadron a New Zealand identity by incorporating the words 'New Zealand' in its official designation — No. 75 (New Zealand) Bomber Squadron.

The beginnings of 75 Squadron go back to the days of the Royal Flying Corps when it was formed at Goldington, in Bedford, on 1 October 1916 as a home defence unit. Its first equipment consisted of a collection of single-seat and two-seat BE type aircraft but it was later re-equipped with Avros and then Sopwith Camels. Disbanded in Essex in 1919 the squadron was reformed in 1937 as a heavy bomber squadron but in March 1939 it became an operational training unit. The squadron was posted to No. 6 (Training) Group soon after war was declared and on 4 April 1940, its 'number plate' was affixed to the New Zealand Flight of the Royal New Zealand Air Force at Feltwell, Norfolk, a station in No. 3 Group.

The RNZAF had selected Squadron Leader M. W. Buckley to train the ferry crews, after he himself had gained experience on Wellingtons, to form the New Zealand Flight (as it was known) and bring the aircraft out to the Dominion. It was logical, therefore, that Buckley should be given command of the squadron when it developed from the New Zealand Flight in April 1940.

The New Zealand Flight had been established at RAF Station Marham, in Norfolk, on 1 June 1939, as an RNZAF formation. Besides the crew members who had been sent from New Zealand, it included New Zealand pilots and navigators who had been serving with the RAF.*

With Buckley directing the flying training and Squadron Leader C. E. Kay in charge of navigation, the team began preparations for the 13,000 mile ferry flight to New Zealand. In August the flight welcomed two new arrivals — Squadron Leader R. J. Cohen, who had been sent out to command the second ferry flight (Buckley's six Wellingtons were to constitute the first) and Flight Lieutenant I. G. Morrison, who was to be a member of it. But the events of the next few weeks overthrew these plans. When war came the New Zealanders were keen to remain in the United Kingdom and engage the enemy.

Buckley was a fine squadron commander, viewed by his men with respect and affection. Alan Mitchell, in his *New Zealanders in the Air War*, wrote: 'I have yet to meet a commanding officer who was more beloved by his squadron.' No. 75 Squadron formed up at Feltwell after brief sojourns (as the New Zealand Flight) at Harwell and Stradishall. More men arrived and the squadron, equipped with Vickers Wellington Mk I aircraft, went into action at once with leaflet raids on Germany. Enemy searchlights were active, turbulence was troublesome and freezing was severe but all the crews returned safely from these missions.

Morale was high and when the Norwegian campaign opened the squadron had its first taste of bombing operations. One of the more interesting missions fell to Flight Lieutenant Breckon. On 12 April 1940 he was detailed to undertake a highly secret reconnaissance of Narvik Fiord, well to the north in Norway. The Germans were present in force in that area besides occupying the key airfields of Stavanger and Aalborg. A naval engagement had just been concluded and Royal Navy units were to sail up the fiord on 13 April to complete the destruction of German units in the Narvik area. Breckon's mission was to reconnoitre the area before the second assault.

With long-range tanks fitted to his Wellington, Breckon took off at dawn from an airfield in the north of Scotland. With him were Pilot Officer D. J. Harkness and E. P. Williams, now a sergeant, as crew, besides a naval observer. In steadily deteriorating weather they flew over 1000 miles along the Norwegian coast to Narvik Fiord. Breckon's airmanship was severely tested during the actual reconnaissance, strong winds and snowstorms

* By late July that year Buckley had a strong team which included Squadron Leader C. E. Kay, Flight Lieutenant C. C. Hunter, Flying Officers J. Adams, A. A. N. Breckon, J. N. Collins, A. B. Greenaway and F. J. Lucas, Pilot Officers H. W. Coleman, T. O. Freeman, W. M. C. Williams, and N. Williams, Sergeants W. D. Steven and T. R. Read and Leading Aircraftmen D. C. McGlashan, J. T. White, E. P. Williams and R. A. J. Anderson.

making control of the bomber difficult. At times the cloud was down almost to sea level. Visibility was reduced to 500 yards or less and the crew members were subjected to violent turbulence. Breckon said later that at one stage, when he was flying at 200 feet in a severe snowstorm, he had doubted that they would find their way out of the fiord in safety. Altogether they were 14 hours in the air, 14 hours of extreme vigilance and attention to the details of flying. The Allies captured Narvik on 28 May.

On the night of 15 May 1940 six Wellingtons of 75 Squadron took part in the first major bombing raid on Germany, against oil installations in the Ruhr, 80 aircraft of Bomber Command being launched against the enemy's industrial heart. No. 75 Squadron was also part of the raiding force which made the first large-scale attack on Berlin, on the night of 23 September 1940, when more than 100 aircraft attacked the German capital as a reprisal for Battle of Britain offensives.

A night mission a few weeks earlier, when the squadron sent 11 aircraft to attack Cologne, was an eventful one for Flying Officer D. L. Prichard and his crew. The starboard engine was hit by enemy fire, the bomb doors were blown open and both front and rear gun turrets were damaged by fire at close range from a Junkers 88. Defenceless now,

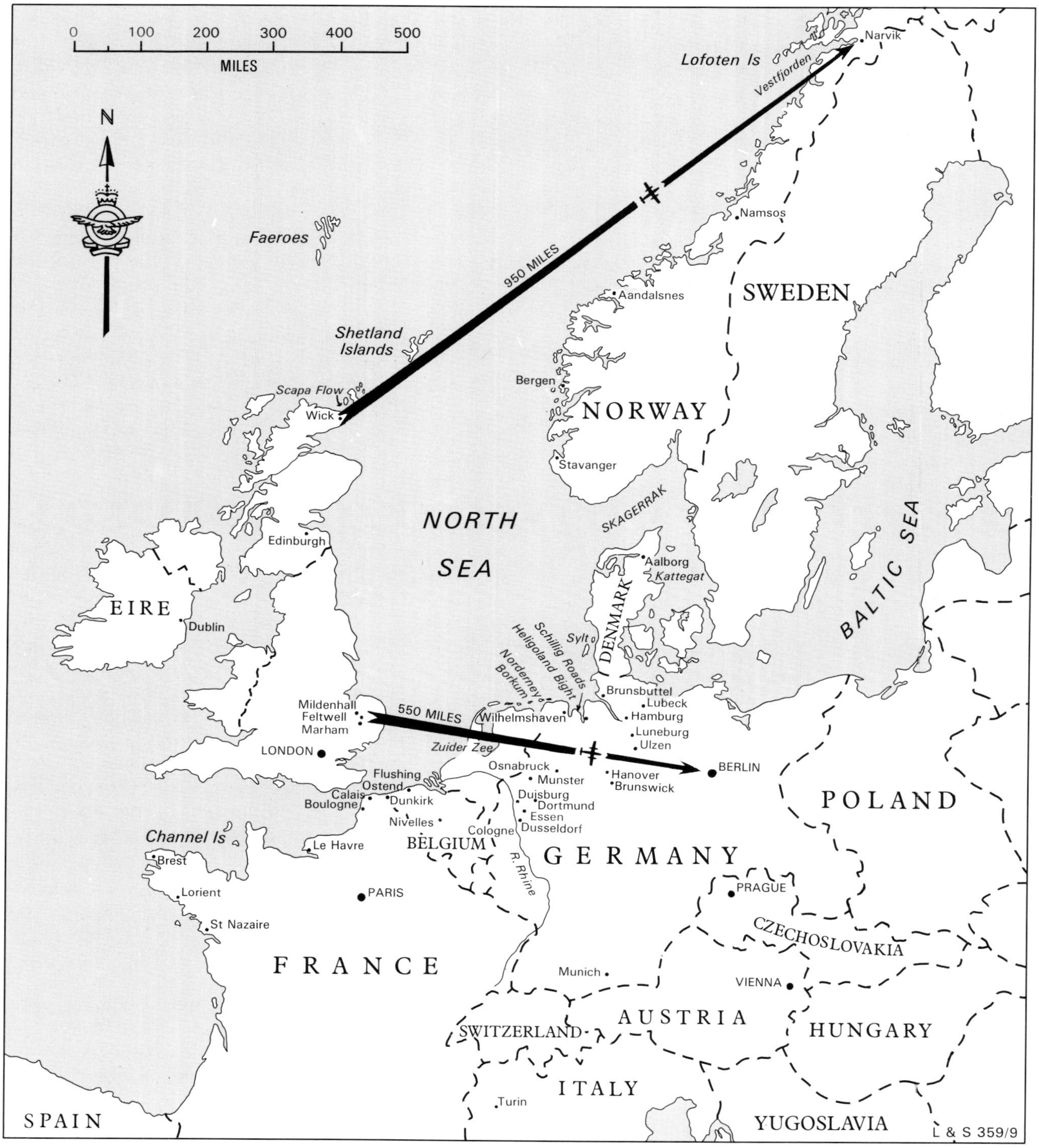

Flight Lieutenant A. A. N. Breckon (second from left) and the crew of a No. 75 Squadron Wellington bomber discuss their secret mission before taking off on a reconnaissance of Narvik Fiord, Norway on 12 April 1940.

and incapable of effective evasive action, Prichard completed his mission, penetrating the city's defences and dropping his bombs in the middle of the target area. He then flew the crippled bomber home to England. It was his fiftieth sortie and it ended with a crash landing at base. The DSO awarded him after this action was the first won for 75 Squadron.

By October 1940 the squadron under Buckley's leadership — he was then a wing commander — was battle seasoned and confident, rapidly winning a place as one of the foremost squadrons in 3 Group. Over the next 18 months the exploits of its aircrews and the high servicing standards of its ground staff were to enhance this reputation and build a tradition unbettered by any other bomber unit.

One of the most successful pilots in 75 Squadron's early days was Trevor Freeman, rated as one of the most experienced navigators in the RAF. He was one of the very few service pilots to hold the No. 1 Civil Navigator's Licence, and for a period, while he was resting from operations, he was in charge of navigational training at No. 3 Group under Wing Commander Andrew McKee* who had left New Zealand to join the RAF in 1926. Freeman was awarded the DFC while he was serving with 75 Squadron and he subsequently won the DSO and a bar to the DFC. After his second tour of operations, when he had completed over 60 missions, he returned to New Zealand and went on to operations in the Pacific.

Within the squadron, each aircrew had its own aircraft and its own (United Kingdom) ground crew, and the names bestowed on the machines and the insignia which accompanied them created a certain individuality and a good-natured rivalry. The aircraft were usually as personal to a fitter or mechanic as to a pilot or navigator. The ground crews were always keen to see their own particular aircraft return from a mission, hear a first-hand account of it, and share in a victory; they were equally keen to repair damage or check for defects. The Wellington bombers carried a crew of six — captain, second pilot, navigator/bomb aimer, wireless operator and front and rear gunners. Of geodetic construction and covered with fabric, the Wellington was the brainchild of Barnes Wallis, the brilliant inventor and designer at the Vickers works at Weybridge. The Wellington was the mainstay of Bomber Command for the first three years of the war.

Munich, Berlin and Hamburg were key targets in November 1940 and, with the knowledge of what Britain had suffered at the hands of Germany, particularly in July and August, 75 Squadron pressed home its attacks with determination. November also brought a change in command. Buckley handed over on 25 November to Cyril Kay, who had been promoted to wing commander, and was himself appointed Commanding Officer of the Feltwell Station, which meant that he did not sever all connections with his old squadron.

There were heavy casualties in the months ahead and some high drama for the men who did get home. On one raid against Hanover, when 11 aircraft from 75 Squadron were part of an attacking force of 220 bombers, the Wellington flown by Flight Lieutenant J. E. S. Morton was three times intercepted on the return journey by enemy night fighters. Cannon shells burst inside the bomb bay, puncturing the hydraulic pipelines so that the bomb doors fell open and the undercarriage hung down, reducing speed and causing the aircraft to lose height whenever evasive action was taken. Despite this, Morton was able to reach a landing field in East Anglia and crash land without injury to any of the crew.

On a raid over Berlin Flight Lieutenant T. F. Gill* was forced down low when over the city and held by a battery of searchlights. His aircraft was repeatedly hit by flak and when he was finally clear he found he was some distance east of Berlin. The navigator had lost his bearings and fuel was getting low. Only Gill's expert airmanship brought the aircraft and crew to a safe landing at base with near-empty fuel tanks. That incident occurred on 23 March 1941. March was an eventful month for Gill. On the 12th his aircraft was attacked by a Junkers 88 during the outward flight to Berlin. His rear gunner opened fire at close range and the German aircraft was seen to stall and dive seawards. In the engagement the Wellington was extensively damaged, so much so that it became difficult to manage, but despite this handicap Flight Lieutenant Gill pressed on to the target and completed his assigned mission. Having flown in the early campaign in France before joining 75 Squadron, Gill was, despite his youth, classed among the veterans. By the age of 23 he had been awarded the DSO.

* Air Marshal Sir Andrew McKee, KCB, CBE, DSO, DFC, AFC, RAF.

* Air Commodore T. F. Gill, CBE, DSO, RNZAF. He entered politics after leaving the Service, becoming MP for East Coast Bays. He subsequently held the portfolio of Minister of Defence and was appointed New Zealand Ambassador to the United States in August 1980. He died in office in America on 1 March 1982.

One of the most colourful members of 75 Squadron was Squadron Leader F. J. Lucas, then and since known universally as 'Popeye'. During a night attack on Dusseldorf, Lucas, then a flight commander of the squadron, dropped his bombs and turned back over the target for a photographic run. A direct hit damaged one engine, the undercarriage dropped and would not retract, and the bomb doors refused to close. On the long and difficult return flight the Wellington steadily lost altitude. Lucas crossed the English coast at 500 feet with petrol gauges showing almost empty. He was directed to a nearby airfield where, as his aircraft rolled off the flarepath, the engines died.

Squadron Leader F. J. Lucas, DFC.

Three New Zealand airmen in the Second World War won the Victoria Cross for outstanding valour in air operations. The first of these awards, to former Wanganui school teacher Sergeant J. A. Ward, followed a large-scale bombing raid on Munster on the night of 7 July 1941, in which 12 aircraft of 75 Squadron took part.

Ward, who was 22 and had been in the RNZAF for almost a year, was flying as second pilot in a Wellington captained by Squadron Leader R. P. Widdowson, RAF, a Canadian. It was a mixed crew. The navigator, Sergeant L. A. Lawton, was RNZAF and came from Auckland; the wireless operator, Sergeant W. H. Mason, was RAF and an Englishman; the front gunner, Sergeant D. H. V. Evans, was RAF and a Welshman; and the rear gunner, Sergeant A. R. J. Box, was RNZAF and came from Auckland. Box, who was only 19, had enlisted in the RNZAF 14 months before.

Widdowson had successfully completed his bombing run and was returning over the Zuider Zee when a Messerschmitt 110 appeared. In the first exchanges, the Welshman was wounded, but Box, from the rear turret, got in a sustained burst from point-blank range and a jubilant crew watched the Messerschmitt falling into the sea with both engines on fire.

The next step was to take stock of their own position — and it was far from reassuring. The starboard engine was vibrating badly; the hydraulic system had been hit and the bomb doors hung downwards; both the intercom and the wireless sets were out of action. Worse, fire had broken out on the starboard wing and was being fed by petrol escaping from a fractured fuel pipe.

'Prepare to abandon aircraft.' With no intercom, Widdowson relayed the warning order to all members of the crew, who could not take their eyes away from the wing section where the fire showed no signs of abating. They hacked a hole in the fuselage with an axe, and brought the portable fire extinguishers into play, but the liquid vanished uselessly in the slipstream before it could reach the fire. It was then that Ward announced quietly that he was going out onto the wing.

Ignoring the protests of his colleagues, Sergeant Ward climbed up through the narrow astro-hatch, dragging with him a loose engine cover. Realising that Ward had made up his mind, Lawton threw a rope around his friend's waist and helped him adjust his parachute. Then Ward was outside, inching his way along the wing, kicking holes in the fabric to obtain a precarious hold, while the slipstream tore at him with the force of a hurricane.

Then he was there, only to have the engine cover plucked from his grasp by the wind and lost in the night. Not to be outdone by the elements after getting so far, Ward beat the flames with his arms and hands, subdued the worst of the fire and started to clamber slowly and painfully back, exhausted beyond description. The return was a nightmare, but the navigator kept a strain on the rope as Ward slowly pulled himself along the wing and up the side of the fuselage to the astro-hatch. He was in.

There was now small danger of the fire spreading and it finally burned itself out. The crew returned to their tasks, Widdowson flying the crippled bomber home with exceptional skill.

Back at base, Wing Commander Kay and his staff officers had received an interrupted garbled message that the aircraft was in some sort of trouble; fire had been mentioned. Then followed anxious hours of waiting with no word from the aircraft until at last, early next morning, advice came through that the Wellington had landed, badly damaged, on the emergency airfield at Newmarket and that the crew would be returning by road.

Their 'Wimpey', as the Wellington was affectionately known, did not fly again. A few weeks later Ward was killed on air operations.

No. 75 Squadron Stirling bombers and their crews before a raid over Germany. *Air Ministry*

Command passed from Kay to Wing Commander R. Sawrey-Cookson, an Englishman, in September 1941, and through the winter of 1941–42 the squadron was active, hitting hard at key targets, including Essen, Kiel, Lubeck, Cologne, Hamburg and Dortmund. In April 1942 Sawrey-Cookson was killed in an attack against Cologne. The unit had been suffering heavy casualties about this time and the death of the CO, coupled with the prevailing loss rate, had a depressing effect on the squadron as a whole.

But Wing Commander E. G. Olson, who succeeded Sawrey-Cookson, soon had morale soaring again. Olson had progressed from his CFI appointments in New Zealand to Air Member for Personnel at the Air Department in Wellington and then to New Zealand Liaison Officer at the Air Ministry in London. His transfer to the command of 75 was both popular and timely. With his flight commanders, Squadron Leader F. Denton, now on his second tour with 75 Squadron, and Squadron Leader R. J. Newton, Wing Commander Olson restored 75's former confidence, performance and spirit. For three months during the following summer the squadron contributed more aircraft to the overall bomber offensive than any other squadron in its group.

Wing Commander E. G. Olson, No. 75 Squadron's fourth CO.

Sergeant J. A. Ward, VC.

In May 1942, the first thousand-bomber raids made world headlines. During this period 75 Squadron operated on five successive nights, participating in the large-scale attacks on Essen and Cologne.

When Olson was promoted to group captain in July that year and given a station (and subsequently awarded the DSO), Wing Commander V. Mitchell, a Scot, took command. He quickly established himself with the New Zealanders and under his leadership the unit maintained the high operational standards that had prevailed during Olson's term. In August 1942 the squadron moved from Feltwell to Mildenhall in Suffolk.

The Wellingtons were now becoming obsolescent for night-bombing operations and the squadron converted to the four-engined Stirling bomber in November 1942, using the new aircraft for the first time in an attack on Turin.

Colour plate: Vickers Wellingtons of No. 75 Squadron crossing the Dutch coast following a raid on enemy targets in occupied Europe.

Lancasters of No. 75 (New Zealand) Bomber Squadron in pre-invasion daylight raids against Channel ports, 1944.

Four-engined Stirling bombers of No. 75 Squadron taking off for a raid on Berlin, March 1943.

C AA

Airspeed Oxfords of No. 1 Service Flying Training School, Wigram.

DH Tiger Moths at an Elementary Flying Training School.

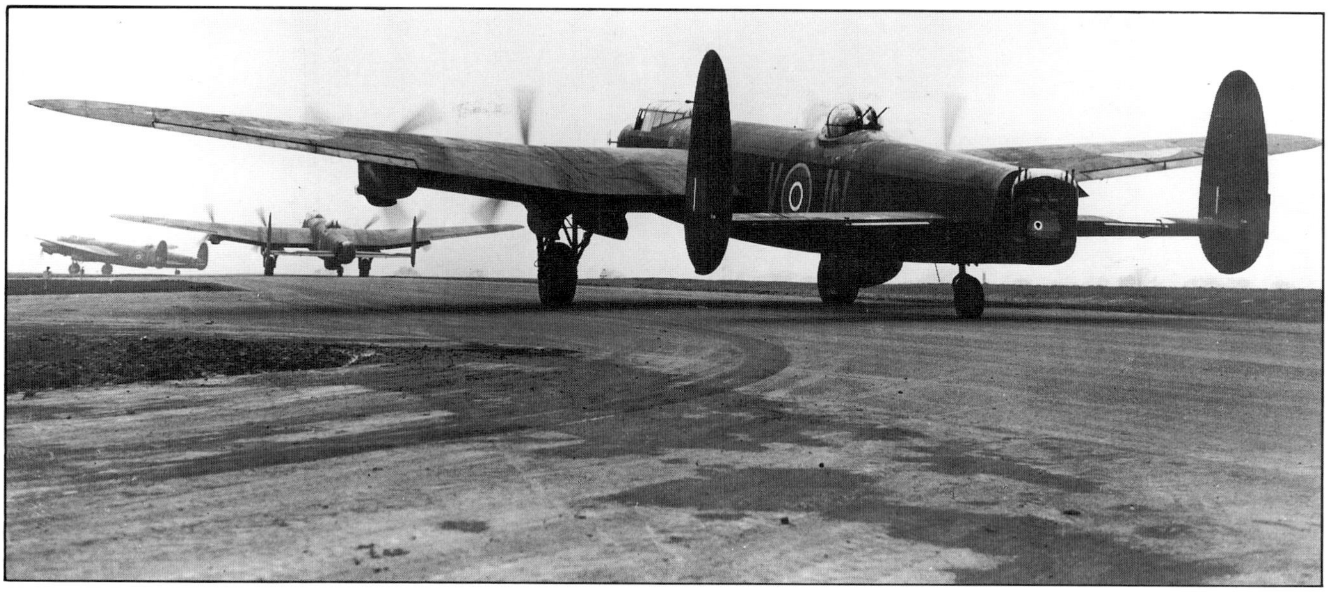

Lancasters of No. 75 Squadron taxi out for take-off on a mission.

The squadron was next commanded by two Englishmen in succession, Wing Commander G. E. Lane, DFC, and Wing Commander M. Wyatt, DFC, both experienced and able, before another New Zealander, Wing Commander R. D. Max, DFC, took the reins. In March 1944 the squadron converted to the Avro Lancaster bomber, with which its crews were more than pleased.

Max, who had won the DSO, was succeeded by Wing Commander R. J. Leslie, AFC, who was followed by Wing Commander R. J. Newton, DFC, and finally by Wing Commander C. H. Baigent, when Newton failed to return from a mission. No. 75 Squadron flew its last bombing raid of the war on 24 April 1945, when 19 aircraft attacked the marshalling yards at Bad Oldesloe.

The following evening a solitary Lancaster captained by Flight Lieutenant H. W. Hooper, RNZAF, flew to The Hague on a peaceful mission — the dropping of leaflets which contained a public apology from the British Government for the wrongful bombing of that city by Allied aircraft. The drop, which was radar directed, was made at 20,000 feet. At the briefing Wing Commander Baigent told the crew that lots had been drawn within Bomber Command to select first the group, then the squadron and finally the crew which would undertake this mission.

Flight Lieutenant H. W. Hooper, the Lancaster captain who dropped leaflets of apology on The Hague in April 1945.

Over five years, 75 Squadron was represented in virtually every major operation controlled by RAF Bomber Command. It took part in the historic attacks on the battleships *Scharnhorst* and *Gneisenau* and in the bombing of German submarine bases. Time and again it topped all the squadrons in 3 Group for successful sorties against the enemy. During its life as a Bomber Command squadron, 75 Squadron flew 8150 sorties, dropped 21,630 tons of bombs and 2344 mines and destroyed at least 45 enemy fighters. Squadron decorations totalled 118 and included one Victoria Cross, six Distinguished Service Orders, 88 Distinguished Flying Crosses, four bars to DFCs, 17 Distinguished Flying Medals and two Conspicuous Gallantry Medals.

Following the defeat of Nazi Germany the Air Officer Commanding No. 3 Group, Air Vice-Marshal R. Harrison, CB, CBE, AFC, RAF, paid this forthright tribute: 'No. 75 (New Zealand) Bomber Squadron has maintained throughout the war an outstanding record unsurpassed by any squadron in Bomber Command.' No fewer than 1370 New Zealand aircrew served with 75 Squadron at some stage of the war. Of them, 422 lost their lives while serving with the unit and another 77 died after transferring to other units.

No. 75 Squadron was disbanded in October 1945, but neither the RAF nor the RNZAF was willing to permit a squadron of such distinction to lapse into oblivion. By agreement between the two air forces, 75 Squadron was recreated in New Zealand, at Ohakea on 11 October 1946, when No. 2 (BR) Squadron was designated No. 75 Squadron.

A No. 488 Squadron Hurricane fighter over Palembang after the fall of Singapore.

CHAPTER SEVEN

The RNZAF in South-East Asia, 1941–42

When war with Japan broke out on 7 December 1941 more than 400 members of the RNZAF were serving in Singapore, Malaya and Burma. Most of the pilots of Nos 67 and 243 Squadrons, Royal Air Force, were members of the RNZAF attached to the RAF. No. 488 (New Zealand) Squadron was drawn almost entirely from the RNZAF and No. 1 (Aerodrome Construction) Squadron, a relatively large unit stationed in Southern Malaya, was manned wholly by RNZAF personnel. There were also RNZAF men in flying-boat and bomber squadrons in the area and in medical, technical and administrative posts in other units.

The attack on Malaya was, in fact, Japan's first onslaught against the Allies, the massive carrier-plane attack on Pearl Harbour coming a little more than an hour later. At about the same time Japan struck at Thailand, the Philippines, Hong Kong, Guam and Wake Island. On this day and on the succeeding days of Japan's south-eastward thrusts, Britain and the United States suffered appalling reverses.

Japan's invasion of Malaya was not entirely unexpected. It had long been apparent that she had ambitions to become the controlling power in South-East Asia and was merely awaiting the opportunity. Colonel Bettington, in his 1919 report to the New Zealand Government, had foreseen the rise of militarism in Japan and had pointed to her growing population and military might as indications that she might one day pose a threat in the Pacific. But the blow which fell in December 1941 was swift and sudden — and it was remarkably successful. Within 70 drama-packed days the entire Malay peninsula, Britain's richest tropical possession, the world's chief source of rubber and one of its principal sources of tin, was in enemy hands. Two mighty British warships, *Prince of Wales* and *Repulse*, were at the bottom of the sea and Singapore and its great naval base (long described as 'an impregnable fortress') had fallen after a brief siege. A shocked British Empire learned that 70,000 exhausted defenders had passed into captivity, a captivity so rigorous that at least half of them would die.

This military debacle, a disaster as great as any suffered by British arms, has been ascribed to a combination of circumstances. There was the climate, for one thing, and the locals' apathy towards the defence of the territory. British troops, though superior to the enemy in numbers, were not trained in jungle warfare. Also, the proficiency of the Japanese in aerial warfare was greatly underestimated. These were all contributing factors, but behind them was the undeniable fact that the plans and preparations made by Great Britain, Australia and New Zealand in the years before the outbreak of war were quite unequal to the situation which developed in 1941.

These plans and preparations had been based on sea power and were therefore concerned principally with the defence of Britain's naval base at Singapore, built at a cost of £65,000,000. For 20 years the view was that any attack on the base would come from warships, supported by carrier-borne aircraft. The possibility that an assault might come from any other direction appears to have been largely discounted. Powerful batteries of fixed guns were in position at Singapore to defend the naval base but they had no forts or fixed defences to protect their rear. There were four bomber squadrons to supplement the gun emplacements, two of which were torpedo-bomber units, and there were two flying-boat squadrons. Their presence, though, was seen as of secondary importance; the defensive plan was essentially a naval one.

This situation continued during the opening months of the Second World War. The collapse of France and the occupation by the Japanese of French Indo-China brought about a rude awakening. Malaya was now likely to be threatened but because of the delicate balance of naval strength in European waters Britain could not send more warships to the Far East. It was decided that the defence of Malaya must depend primarily on air power, supported by such land forces as could be made available. This was at the end of July 1940 when there were precisely 84 operational aircraft in the whole of Malaya, not one of which was a modern fighter! Plans were co-ordinated between Britain, Australia and New Zealand for substantial reinforcements (including trained pilots) to be sent to the Far East but in the conditions prevailing in 1940 such things could not be done in a matter of weeks nor even of months. For Britain herself was in grave danger; the Battle of Britain had begun. The result was that when the Japanese launched their attack these plans were nothing like fulfilled. For example, instead of 582 first-line aircraft (the number considered necessary to meet the changed situation), there were only 158, many of them obsolete.

To support her invasion of Malaya, Japan had a force of 300 land-based aircraft deployed in Indo-China in addition to her carrier-based squadrons. They included the twin-engined Army types 97 and 99 for bombing and reconnaissance and the Navy type 96 for use as a torpedo-bomber. Her principal fighter aircraft were the Army types 1 and 97 and the Navy type 0. The last named, better known as the Zero,* proved one of the greatest surprises of the campaign. Fast, manoeuvrable and well armed, it was more than a match for our fighters.

To oppose the formidable Japanese forces Britain had, by November of 1941, a fighter defence of four units — No. 243 Squadron, RAF, No. 21 Squadron, RAAF, No. 453 Squadron, RAAF, and No. 488 (NZ) Squadron, RAF. Between them, these four squadrons had a total of 43 Buffaloes. In addition there were in Malaya and Singapore two RAF squadrons of Blenheim bombers, one RAF squadron of Blenheim night-fighters, two RAF squadrons of Vildebeest torpedo-bombers, two Australian squadrons of Hudson bombers, one RAF general reconnaissance squadron with three Catalina flying-boats, one RAF bomber squadron of Mark IV Blenheims, and an anti-aircraft co-operation unit with 12 assorted aircraft. There were also on Singapore Island three Catalina flying-boats manned by Dutch crews.

The existence of No. 488 (NZ) Squadron requires some explanation. The agreement with Britain and Australia to make a contribution towards bolstering British defences in Malaya led the New Zealand War Cabinet on 25 June 1941 to authorise the formation of a New Zealand squadron for Far East service with the RAF under the terms of Article 15 of the Empire Air Training Scheme Agreement. Initially the approved establishment was nine officers, 11 sergeant-pilots and 132 ground staff. (The pilot strength was later increased to 26.) The squadron was to be raised in New Zealand and moved with all speed to the Far East. The RAF named this new unit No. 488 (NZ) Squadron.

The RAF provided the squadron commander and the two flight commanders by transferring them from RAF units. They were all New Zealanders. Squadron Leader W. G. Clouston, DFC, a veteran of the Battle of Britain, was serving in the RAF. He was 25 years of age when he arrived in Singapore on 19 September 1941 to take command of 488 Squadron. Flight Lieutenant J. N. MacKenzie, DFC, was also serving in the RAF and had fought in the Battle of Britain. He was 27. (He subsequently transferred to the RNZAF.) Flight Lieutenant J. R. Hutcheson, the other flight commander, was an RNZAF officer who had been serving since 1939 and he had been flying in a Hurricane squadron commanded by Clouston. At 29, Hutcheson was the oldest of the three.

Clouston was joined by his flight commanders on 29 September. They still had no squadron. At RAF Station Kallang, close to the city of Singapore, they spent their days preparing for the arrival of the main body of pilots and ground crew from New Zealand and getting acquainted with the Brewster Buffaloes they were to fly. 'A nice old gentleman's aeroplane', was MacKenzie's cryptic comment after his first flight on 30 September.

They were joined on 5 October by an advance party of three airmen, Leading Aircraftmen K. F. Burnell and R. W. Gillatt and Aircraftman E. C. Hooper. Five days later the six of them went to the Singapore docks to welcome the main body which arrived aboard the Dutch passenger ship *Tasman*. Another party was to arrive in November so the squadron could not be fully operational until then.

At Kallang, which was to be their home for a few short but eventful months, Clouston lost no time in getting training under way. Most of the squadron had received little preparation in New Zealand for the ordeal which awaited them. The airmen had assembled at RNZAF Station Rongotai (Wellington) in September and undergone drill and weapon training, but the pilots had flown only Harvard trainers and were without experience in air weapons.

There was an operational training unit at Kluang in Johore and Clouston had his pilots there within 48 hours of their arrival to begin conversion on to Buffaloes. The American-built Buffalo was a single-engined fighter powered with a Wright Cyclone radial engine. It had a cruising speed of about 220 mph and a cruising range of about 1000 miles. Sadly, the New Zealanders were to find that in rate of climb, ceiling and manoeuvrability the Buffalo was vastly inferior to the Japanese fighters.

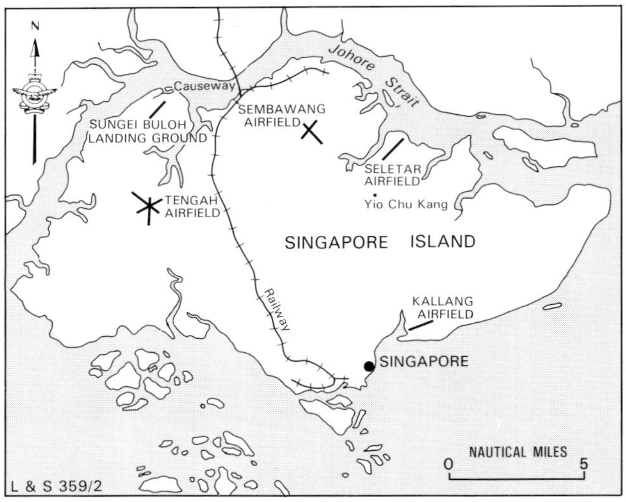

At Kallang the ground element of 488 Squadron took over the 21 aircraft left behind by No. 67 Squadron, RAF, which had been transferred to Burma. Some rude shocks awaited them. First, the Buffaloes were in various stages of disrepair. Second, the departing squadron had taken with them all the tools, spares and accessories. Only the most perfunctory of inventory checks was necessary to reveal the grim state of affairs: the equipment bequeathed to 488 Squadron comprised six trestles, six chocks, one broken ladder and six oil-draining drums! And a great deal of work would have to be done on the aircraft to ready them for operational flying.

There was no evidence that the RAF at Kallang had made any preparations for their coming. Flying Officer Cecil Franks, the squadron equipment officer, had to begin from scratch to build up supplies of materials for aircraft maintenance. In the supply organisation at Kallang the effects of the red tape of peacetime administration were obvious and the acquisition of stores through official channels was painfully slow. Franks, however, was never lacking in initiative and he soon established a direct liaison with the equipment depot at RAF Station Seletar. His friendly dealings with the depot not only furnished his

* The Mitsubishi Type O or Zero bore the Allied codename 'Zeke'.

own squadron with essential supplies in a remarkably short time but also helped to equip No. 243 Squadron, RAF, also based at Kallang. When the 488 Squadron pilots returned from their conversion training, their Buffaloes were ready to fly, but at the expense of gruelling hard work and infinite patience on the part of the RNZAF maintenance staff.

Although the pilots were deemed to have converted on to Buffaloes at Kluang, in fact they had been given only a few hours' refresher flying and this had been done on Wirraway aircraft, an Australian version of the Harvard trainer. At Kallang, too, they had to start from scratch, progressing from 'circuits and bumps' to aerobatics and, once they were proficient in pure flying, to operational exercises — map reading, reconnaissance, Army co-operation, formation flying and combat tactics.

ibly, the routine was still that of peacetime, with a half-day holiday every Wednesday and no work on Sundays. Yet over the island hung the grim threat of war.

The New Zealanders were not accustomed to the leisurely ways of the Englishman abroad and it seemed to them that too much time was being frittered away. Another real problem was the acute shortage of ground staff, particularly with the volume of work demanded by old and decrepit aircraft. Additional men were attached to 488 from other Singapore-based units — after some table banging by Clouston — but the New Zealand servicing crews were still overworked.

Fortunately they were well led and constantly inspired by a man with all the attributes of true leadership, Flight Sergeant W. A. Chandler. At Kallang Chandler put his men to work modifying and repairing the Buffaloes. He had a good stalwart in Sergeant Stan Guineven who, in an unobtrusive way, got through a great volume of work. Because such a high proportion of the aircraft were always serviceable the squadron soon achieved a standard of operational efficiency few would have believed possible.

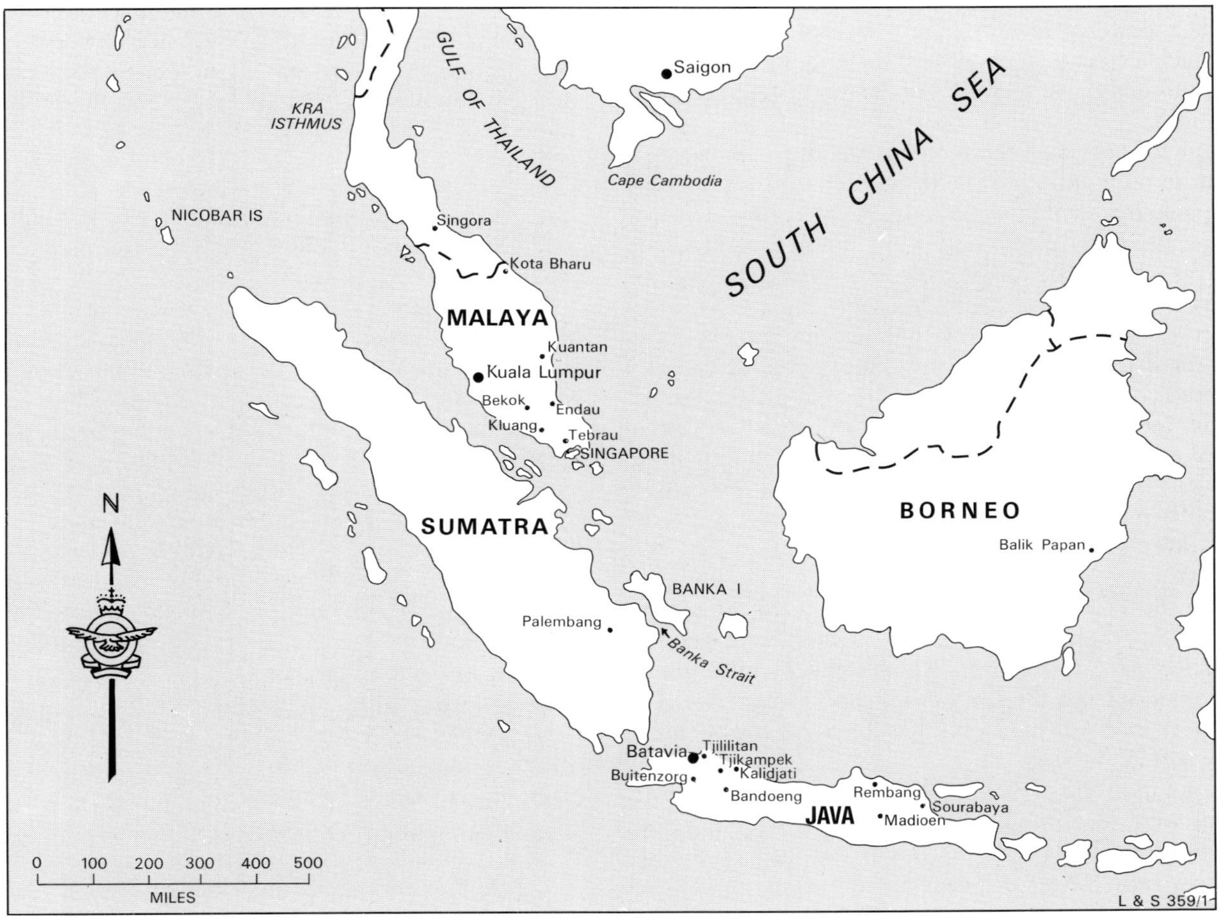

With Clouston almost fully occupied in administration of the squadron, MacKenzie and Hutcheson shouldered most of the training. The load was eased in December when Pilot Officers G. L. Hesketh and F. W. J. Oakden were posted to the squadron as assistant flight commanders on loan from 243 Squadron.

Clouston's goal was a squadron of top proficiency, but there were two militating influences. One was the weather — every afternoon heavy thundershowers, part of the weather pattern in those regions, grounded aircraft temporarily and stopped flying. The other was the extraordinary indifference to the urgency of the situation manifest in the higher command in Singapore. Incredibly ...

Tension was already high when on 1 December General Headquarters Malaya ordered second-degree readiness, which meant that all forces were to be ready for operations at short notice. On that day the air raid alert system came into operation. Then reports filtered down that the Japanese were in fact moving southward and an air reconnaissance of the waters to the east and north of Malaya was ordered.

At noon on 6 December a Hudson of No. 1 Squadron, RAAF, sighted a large Japanese convoy in the South China Sea 205 miles from Kota Bharu in northern Malaya. The convoy was moving on a westerly course towards Malaya. Shortly after midnight on 8 December this convoy landed Japanese troops at Kota Bharu and the Australians operating out of the rough airfield there were among the first Allied airmen to strike from the air against the Japanese in the Second World War.

No. 1 (Aerodrome Construction) Squadron, the first unit of its kind in the air forces of the British Empire, had been formed in New Zealand in July 1941 and had reached its full strength of 15 officers and 140 other ranks in October. Recruited from the then Public Works Department and from privately owned construction companies as well as from the ranks of the RNZAF itself, this squadron was made up of men somewhat older than those in other air force sections, their average age being 36. They had been chosen because of their ability to stand up to continuous hard work under gruelling tropical conditions.

From its base at Tebrau in Johore the Construction Squadron built an airfield, with two runways forming an L, suitable for bomber operations. First they had to clear the area of rubber plantations, their bulldozers uprooting the trees which were thrust aside to be disposed of by the hundreds of locally--engaged labourers employed on the project. The north-east monsoons had started when the squadron began the task, and every afternoon, with clock-like precision, the rains fell, turning the newly cleared ground into soft mud. Tractors and carryalls became bogged to their axles and it was impossible to put them to work. Inevitably, this meant waiting for the next day's sun to dry out the ground. The unit made up for lost time by working around the clock whenever a prolonged dry spell permitted.

Once the Tebrau airport was well advanced the officer commanding, Squadron Leader E. C. Smart, arranged for the survey party to move to Bekok, some 90 miles to the north, to mark out the site for a second bomber airfield. This party made the journey under the command of Flight Lieutenant A. C. Begg.

On the morning of 8 December the New Zealanders at Tebrau were awakened early by a sound they were to remember for a long time. Johore is just over the causeway from Singapore Island and the camp was not too many miles distant. It was the air raid sirens of Singapore that disturbed their sleep and when the men ran from their huts into the moonlight they found they had a grandstand view of Japan's first air raid on Singapore. There were brilliant flashes of light as bomb after bomb exploded on Singapore Island and the tracer from ground defences was clearly visible as it sought out the attacking aircraft.

In the face of this raid 488 Squadron flew its first defensive fighter sortie. Flight Lieutenant John Hutcheson was duty officer at Kallang on the night of 7 December and it was he who first received the news that Japan had landed troops on the Malayan beaches. Enemy bombers were over Singapore by 4 a.m. attacking Tengah and Seletar airfields. They had been picked up by radar some 130 miles away but because of a breakdown in communications the lights of Singapore had not been extinguished and there was no blackout. Hutcheson lost no time in getting an airborne patrol up and setting a pattern for what was to be an arduous day.

Early on the morning of 10 December 488 Squadron was told that it might be required to give cover to two warships. No names were given. Two aircraft were to take off every 30 minutes, fly to a given patrol area, stay there for half an hour and return. The squadron heard later that morning that the battleships *Prince of Wales* and *Repulse* were being attacked by enemy aircraft some 170 miles from Singapore. Both ships had been at Singapore when war broke out and were returning after attempts to intercept enemy convoys in the Gulf of Siam when they were attacked. The Japanese aircraft had sighted the vessels when they emerged from protective cloud layers into clear weather.

When Flight Lieutenant MacKenzie and Sergeant W. J. N. MacIntosh arrived at the scene of action as part of the protective fighter force, both the warships had been sunk. They gave escort instead to a destroyer which had taken survivors from the sea. Other members of 488 Squadron gave air cover over the area where more survivors were being rescued.

On 15 December four members of the squadron led by MacKenzie pursued a Japanese reconnaissance aircraft. Flight Lieutenant Hutcheson and Sergeant D. L. Clow both sighted the enemy but their Buffalo fighters had neither the height to intercept it nor the speed to prevent its escape. Again and again the Japanese were shown to have superior aircraft and well-trained pilots.

The safe arrival at Singapore of reinforcements was of first importance and although it meant a reduction in support for the Army on the Malayan mainland the principal task for the air forces in the area became the protection of convoys into Singapore. No. 488 Squadron was constantly patrolling the seas with its sister squadrons and giving escort to Allied shipping.

The overall military situation was worsening fast; the British ground forces in Malaya were retreating steadily before the advancing Japanese, and the enemy, occupying more and more aerodromes in northern and central Malaya, was able to increase air activity against Singapore. In January heavy daylight raids against Singapore's airfields began.

Kallang was bombed heavily on 9 January and 488 Squadron's offices and the station's equipment, ammunition and oil stores were almost wiped out. On 10 January the first enemy aircraft to be brought down over Singapore was destroyed by two New Zealanders, Sergeants C. T. Kronk and B. S. Wipiti, both of 243 Squadron.

The first air combat for 488 Squadron came on 12 January. A formation of eight Buffaloes led by Hutcheson and another six aircraft led by MacKenzie were over Johore when MacKenzie's formation sighted about 27 Japanese aircraft, 3000 feet above them. The enemy dived as the Buffaloes took evasive action. Sergeant T. W.

Honan and Sergeant R. W. MacMillan were shot down in a matter of seconds. Both parachuted to safety, Honan with a bullet wound in the left arm. Five other aircraft were damaged and two pilots wounded.

Equally severe on the New Zealanders were the next day's operations. Flight Lieutenant Hutcheson, leading a flight of eight Buffaloes, was ordered to intercept 30 Japanese bombers. Pilot Officer W. J. Greenhalgh attacked an Army 96 bomber and managed to get smoke from one engine. In this engagement 488 Squadron lost four aircraft and had one damaged, but all pilots survived, some with minor injuries.

The squadron had its first combat success and suffered its first battle fatality on 15 January. Hesketh, who led a Buffalo formation on a mission to intercept a Japanese raiding force, was shot down and killed when a swarm of

Pilots of No. 488 Squadron at Kallang, December 1941. From left: Pilot Officer P. D. Gifford, Sergeant J. F. Burton, Pilot Officer F. S. Johnstone and Sergeant E. E. G. Kuhn. *Ministry of Information, Singapore*

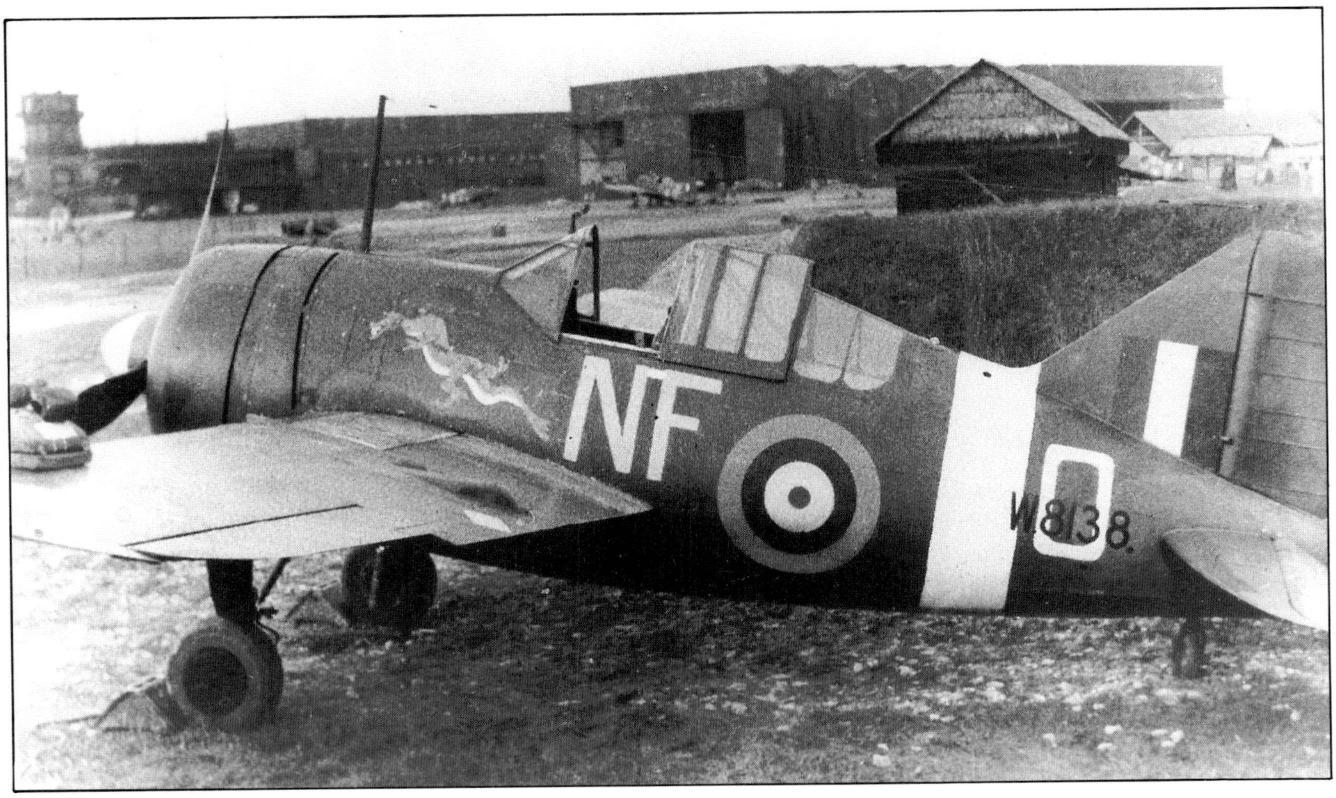

No. 488 Squadron Buffalo at Kallang.

Japanese fighters dived upon them. Sergeant E. E. G. Kuhn, who was to win a DFM and become a prisoner of the Japanese, scored 488's first victory when he shot down a Type 97. Most of Clouston's aircraft were damaged in this action, but all the pilots except Hesketh made base.

One of the squadron's brighter days was 18 January when Hutcheson led a successful attack with support from 243 Squadron. They found themselves engaged with nine Zeros, and were able to report two destroyed and three probables with no loss to themselves. In subsequent engagements later that day and on the following day, however, three pilots of 488 Squadron were lost in enemy action.

The airfield at Kallang was heavily bombed on 22 January. Four aircraft led by MacKenzie were taxiing before take-off when the bombs began to fall. The pilots opened their throttles and raced off but a falling bomb caught one of the aircraft which blasted into a petrol store just after it became airborne. Pilot Officer L. R. Farr died later from injuries. In this same attack two members of the ground staff died at their posts after starting up an aircraft. At this stage the squadron had lost 21 aircraft, most but not all to enemy action, five pilots (one on attachment from 243 Squadron) and two ground staff.

Next day Clouston was rallying his squadron when news came through of his posting to RAF Headquarters, Singapore. He handed the squadron over to MacKenzie, who celebrated his promotion by taking No. 488 into action against a superior enemy force over the mainland.

The squadron then began to re-equip with Hawker Hurricanes which had arrived by ship on 13 January. Nine were allocated to 488. Sadly, all but one of the new aircraft were either destroyed or damaged on 27 January when the Japanese made two heavy raids on the airfield. Twenty-seven enemy bombers came in from the north in the first raid and the aerodrome 'literally rocked for 25 seconds', according to an entry in the squadron diary. MacKenzie and seven pilots were sheltering in a sand-bagged gun emplacement when a bomb fell against the sandbags. The eight pilots were buried when the bomb burst but were later dragged out unhurt; it was a very lucky escape. The second attack came within 40 minutes of the first, with the same number of Japanese aircraft, and left tankers and aircraft burning fiercely on the ground. Squadron members hurried to 243 Squadron to give aid and found that most of that unit's Buffaloes had been destroyed or damaged.

For the next few days everyone worked feverishly repairing the least damaged aircraft and filling in bomb craters so that by 30 January a single strip had been cleared and three Hurricanes were ready to take the air. But the general situation in Singapore had deteriorated badly. The next day the causeway across the Johore Straits was blown up. That evening MacKenzie was told to prepare to take his squadron out of Singapore.

Next day a disillusioned squadron was told that while the other fighter forces and the new planes which had arrived by aircraft carrier were on their way to Sumatra and Java, 488 would not be evacuated yet but would remain to service a new RAF squadron, No. 232, which had just arrived and had no ground staff.

The New Zealanders were to have the distinction of being the last squadron remaining in Singapore with its ground staff. In fact, four of 488's Hurricanes were flown to Palembang in Sumatra on 2 February. The following day Pilot Officer P. D. Gifford and Flight Sergeant J. Rees took a party to Sembawang to service the Hurricanes of the new RAF squadron, arriving amid a salvo of shells.

MacKenzie received his orders at last and sailed for Batavia on 6 February with eight of his pilots. Hutcheson and Oakden, after scouring the island for any serviceable fighter planes that might have been overlooked, followed in the same convoy with the rest of the squadron pilots.

Meanwhile in Singapore the ground crew started servicing those planes which had been flown into Kallang. The Japanese landed on the island on the night of 8 February. Although chances of withdrawal were slender, the New Zealanders gave the RAF aircraft the same care they would have accorded their own. On 10 February the last Hurricane left the island and the New Zealanders contemplated the few other types remaining, battered and in poor condition. With the inspiration provided by Chandler they set to work on these with the thought of producing a serviceable aircraft in which Squadron Leader Clouston might escape. But this was not to be. Clouston was captured when Singapore fell and spent the rest of the war in prison camps.

On the morning of 11 February the 488 ground staff received the welcome news: they were to be evacuated by sea that afternoon. To the Singapore docks they went in good spirits and boarded the *Empire Star*. Two waves of enemy bombers came over the ship as they embarked and the docks took a pounding but at 6.30 p.m. the ship pulled out into the stream and anchored, sailing on the tide 12 hours later for Batavia. Two hours out from Singapore the ship was attacked by successive waves of aircraft. The vessel took three direct hits. New Zealand airmen manned machine guns and some even fired rifles; one Japanese aircraft was shot down and another was damaged. There were some casualties among the ship's passengers, but 488 Squadron came through the ordeal unscathed. Although there were more raids defensive fire kept the aircraft high and there were no more direct hits. The ship reached Batavia's relative calm on 13 February at evening and the men went ashore next morning. They had fought against massive odds to the very last.

Squadron Leader MacKenzie lost no time in reforming his squadron for its new task of defending Batavia from attack. Some Hurricane-experienced tradesmen already on the island were posted to the unit and at Tjililitan aerodrome, 10 miles from Batavia, MacKenzie personally supervised the delivery of his new Hurricanes and established his base. On 14 February Hutcheson was detailed to take nine Hurricanes to Palembang. He, Sergeant H. J. Meharry and Pilot Officer N. C. Sharp and six pilots from 232 Squadron were to leave their aircraft at Palembang as reinforcements and fly back in a transport aircraft. They had no maps and no R/T to cope with the featureless and difficult terrain and the Hurricanes had fuel for the journey and no more. Approaching Palembang, with some of his formation perilously low on fuel, Hutcheson saw the airfield. He also saw little puffs below that he at first assumed to be ack-ack, but then identified as paratroopers descending on the airfield. Two shadows passed swiftly over his vision and he knew his detachment was about to engage the enemy. There were Zero fighters intercepting the formation, Zeros giving cover to the paratroop raid on Palembang.

Although the Hurricane pilots gave fight, they could not stay airborne long after the lengthy flight from Batavia. Sergeant Jack Meharry managed to land despite the warfare on the ground. (He later managed to refuel and flew to a secret base known as P2 from which he flew on operations against the enemy until his Hurricane could fly no more.) The others were either shot down or ran out of fuel. The result in each case was the same; they all finished up in the jungle. Hutcheson and Sharp both survived their crash landings and, with Meharry, were able to rejoin their unit on 17 February.

The bomber strip at Tebrau nearing completion by the men of No. 1 Aerodrome Construction Squadron.

Hopes that southern Sumatra and Java could be held against the Japanese were to fade quickly. Singapore had fallen. Palembang fell to the attackers after the paratroop raid. Total withdrawal from the East Indies was but a matter of time. The Allies' hopes for consolidating their defences in the Indies were frustrated by the incredible speed of the Japanese advance. Java now faced the prospect of invasion and General Wavell, who had moved his Supreme Command headquarters from Singapore to Java a few days before, was to turn his remaining Allied forces over to the Dutch.

Losses in combat had depleted all the squadrons in the area, spares were scarce and disorganisation threatened. On 22 February Squadron Leader MacKenzie received orders to withdraw his squadron to Australia. His aircraft were to be handed over to 605 Squadron RAF and he was to leave with them one flight commander and five other pilots who, it was said, would be evacuated as soon as RAF replacements arrived. Regretfully, MacKenzie and the rest of the squadron said goodbye to Oakden and Sharp and to Pilot Officers H. S. Pettit and G. P. White and Sergeants Kuhn and MacIntosh. Next day 488 sailed for Fremantle.

The RAF was not able to fulfil its promise to evacuate the New Zealand pilots. They fought on until Java surrendered on 8 March. Noel Sharp was shot down behind the Japanese lines and although he was seen to get his plane down successfully, and to climb out, he was not heard of again. The others were taken prisoner. White died in a prison camp and the others survived three and a half years of prison ordeal until Japan's surrender brought their release.

Aboard the liner *Esperance Bay* the tattered but proud remnants of a fine fighting unit returned home in March 1942.

Of 488 Squadron which had arrived in Singapore as a semi-trained, untried unit, Air Vice-Marshal P. C. Maltby, Air Officer Commanding the Royal Air Force in Java, had this to say at the end of the campaign: 'I consider 488 the squadron which has done the best job in Singapore.' Lieutenant General A. E. Percival, GOC Malaya, said in a dispatch: 'I wish to pay tribute to the gallant air crews who throughout the later stages of the Malayan campaign went unflinchingly to almost certain death in obsolete aircraft which should have been replaced years before.'

The story of 488 Squadron's gallant stand is told here in some detail because so many of its personnel were members of the RNZAF, but there were RNZAF aircrew serving in other RAF squadrons in the defence of Singapore who also fought against superior odds. Squadron Leader P. L. D. Cummins, who flew with a Blenheim squadron as a junior officer at that time, said after the war: 'Our handful of Blenheims would be put up as fighter escort for Allied ships assisting in the evacuation and we would meet 54 attacking Japanese aircraft . . . We lost eight aircraft in one day, most of them crewed by young Australians who had no operational experience. The losses were tragic.'

New Zealanders were also prominent with No. 62 (Bomber) Squadron, RAF, which was at Alor Star when Japan invaded Malaya. During December, 18 Hudsons and crews, among whom were six New Zealanders, arrived from Britain as reinforcements. This flight, under the command of Squadron Leader L. G. W. Lilly, arrived early in January and went straight on to operations but, with the enemy moving rapidly southwards, the Hudsons were moved to Sumatra where they flew from the strip known as P2. On 4 February Lilly led a successful raid against Kluang, now held by the Japanese and protected by artillery. When the Japanese invaded Sumatra No. 62 Squadron attacked their ships, on one occasion losing almost half the force the squadron had dispatched. A very successful raid was led by Flying Officer E. J. Henry, RNZAF, with five squadron aircraft on 13 February. They attacked a Japanese convoy, flying into heavy anti-aircraft fire at sea level, sighted three cruisers and three destroyers beside the transports, and scored hits on one destroyer and one cruiser. The Hudsons returned in thunderstorm conditions and nil visibility but the navigation was excellent and four aircraft landed safely at the main Palembang airfield while the fifth put down at P2.

Pilots of No. 488 Squadron outside their dispersal hut at Kallang in late 1941. *New Zealand Herald*

Exhausted members of No. 488 Squadron aboard the *Empire Star* bound for Batavia.

Among the RNZAF personnel who served in Sumatra and Java during the final stages of the campaign were seven pilots of No. 232 Squadron, the unit sent out from England by aircraft carrier to reinforce Singapore. Flight Lieutenants I. Julian and E. C. Gartrell were both awarded the Distinguished Flying Cross for gallantry and leadership during these bitter actions. Julian, who led the final flight to operate from Java, destroyed at least four enemy aircraft in the air and one on the ground while Gartrell also had four definite kills to his credit besides several 'probables'. No. 232 Squadron fought on until the surrender, its members becoming prisoners of war.

It only remains to account for No. 1 (Aerodrome Construction) Squadron. The loss of aerodromes in northern Malaya in the early stages of the war made it imperative to develop new ones in the south as quickly as possible. To achieve this, the development of Tebrau was restricted to the completion of only one runway. The squadron was split up, therefore, and men dispatched to various other jobs. A large detachment was transferred to Bekok with instructions to build a fighter strip there. Another party was sent to Singapore to begin an airfield at Sungei Buloh near Tengah. There were other detachments at Tengah and Seletar and a salvage party in northern Malaya.

The salvage party was responsible for saving enormous quantities of equipment from under the very nose of the enemy. Operating for most of the time only one jump ahead of the British rearguard, they gathered up abandoned lorries, graders and steam rollers, and sent them rolling southward to Singapore in the charge of native drivers. From bombed-out aerodromes they collected priceless loads of radio equipment and sent this, too, with the southbound convoys. All this went on throughout the 500 mile retreat to Singapore.

Gradually the unit reassembled at Tebrau. They were saddened men, who had built airfields with pride in their work, only to abandon them to the advancing enemy, dragging trees across the surface or laying mines for their demolition. Civilian labourers dug holes in the newly formed concrete runways and mines were laid in them ready to be exploded when the command came. 'It would be simpler to build mines into the foundations in future,' one airman remarked bitterly.

The squadron's camp was stripped of all useful equipment and personal gear was packed; on the morning of 27 January — the day 488's Hurricanes were being bombed in a major attack on Kallang field — the squadron moved out. No. 1 (Aerodrome Construction) Squadron was the last air force unit to leave Malaya.

At Singapore they were accommodated at the Dairy Farm. There was plenty for them to do. At first their energies were concentrated on the Sungei Buloh strip and another at Yio Chu Kang, but these projects were soon abandoned in early February when enemy artillery was brought to bear on them.

The need for an aerodrome construction squadron was now being questioned and higher command finally conceded that Singapore now had no use for No. 1 Squadron. The existing fields were being pounded to ruins and any new construction must share the same fate. As in Norway, Greece and Crete, it was being forcibly demonstrated that aerodromes without adequate fighter protection were valueless.

No. 1 Squadron was authorised to depart — destination Sumatra, with a promise of new airfields to build there.

Like 488 Squadron's, its evacuation was not without incident. On the morning of 1 February the squadron was ordered to embark with its machinery on SS *Talthybius*. There was no civilian labour to be found and the ship's crew was untrained in stevedoring, so the squadron members had to load their equipment themselves, operate the winches and stow the cargo. They worked until nightfall and throughout the following day despite air raids. The men stopped to take cover only when enemy aircraft came directly overhead. Several bombs fell nearby but the ship was not hit. In 16 hours' actual working time the New Zealanders loaded nearly 2500 tons of heavy equipment.

That night they returned to their camp. Next morning a working party went to the ship to load some additional RAF equipment. There were two heavy raids during the morning and the *Talthybius* suffered two direct hits, the bombs exploding in the holds. The working party was caught on board; one man was killed and seven others suffered serious injuries, one dying next day in hospital. The ship was set on fire and water poured through the holes blasted in her sides.

After 24 hours of battling to save the ship, the crew brought the fires under control, but she continued to make water in spite of the rigging of auxiliary pumps. Much of the cargo was damaged but the squadron's examination showed that the heavy excavation machinery and medical, dental and other stores were intact. On the afternoon of 4 February a party of squadron volunteers went back to the ship and unloaded the medical supplies and the men's kitbags. It was as well that they did; another bombing attack that afternoon set the ship on fire and this time she sank.

For two days the squadron marked time at the Dairy Farm waiting for orders. On 6 February they went back to the docks, loaded what was left of their equipment into the SS *City of Canterbury* and SS *Darvel* and went aboard the vessels. The *City of Canterbury* reached Batavia under strong naval escort on 9 February.

The *Darvel* was insufficiently crewed and squadron members helped to work the ship. Some took shifts in the engine room, some mounted light anti-aircraft guns and others undertook the messing for the troops on board. A formation of enemy bombers found the *Darvel* in Banka Strait, where she had anchored in the shelter of a group of small islands in the hope of escaping detection. Although there were no direct hits, the ship's hull was riddled with holes from bomb splinters and she was leaking badly. The steering gear was damaged and so were the lifeboats. One New Zealander was killed and at least 17 were wounded in this action.

The order was given to abandon ship, but the boats were unequal to the task. The RNZAF men went below and plugged holes. Others worked on repairing the lifeboats while medical orderlies tended the wounded. An RN officer took over command of the ship and when darkness fell a course was set for Batavia. Next morning the vessel was listing badly to port and the captain announced that the *Darvel* was sinking. Passengers and luggage were concentrated on the starboard side. Members of the Construction Squadron made a final effort at plugging holes. They succeeded after two hours' work; the leaking was brought under control and the *Darvel* eventually reached Batavia on 12 February. Senior service officers who travelled aboard her said that, although the New Zealanders represented only a small proportion of the troops on board, it was due entirely to their efforts and their initiative that the crippled ship reached Java at all.

At Batavia the *Darvel's* detachment rejoined the rest of the squadron and the wounded were treated at the Dutch Military Hospital. Squadron Leader Smart discussed their future role with Allied Air Headquarters at Bandoeng. The military situation was confused and the Japanese were advancing steadily. In one of the last convoys to get away from Java the squadron was evacuated to Australia without further incident. In this manner, No. 1 Squadron came home.

The Fall of Rangoon

In December 1941 only 37 front line aircraft, British and American, were stationed in Burma although the planned strength was a minimum of 280. This force comprised the 16 Brewster Buffaloes of No. 67 Squadron, RAF (which had been transferred from Singapore in October) and 21 Tomahawks of Colonel Claire Chennault's American Volunteer Group, the 'Flying Tigers'. No. 67 Squadron was virtually a New Zealand unit since, with the exception of the officer commanding, his flight commanders and one other, all the pilots were members of the RNZAF. Most of them had gone to Singapore straight from RNZAF flying training schools.

For the Japanese High Command, Burma had much to offer. There was oil, rice and new potential bases which could facilitate an invasion of India. Foremost in the Japanese minds, however, was the urgent need to stem the flow of supplies reaching China via Burma. As Rangoon was the start of the supply line to China and the main port for the Allied defence of Burma, the Japanese began their invasion by bombing Rangoon and then attacking towards that city.

Burma's defences were not great. There were 25,000 combat troops on the ground, mostly local recruits, only partly trained and not well equipped for the role they were expected to undertake. The main defensive line was the River Salween and to help the Army hold it plans had been drawn up for the construction of eight airfields. Only seven had been built when war came and they extended from Lashio in the north to Mingaladon in the south. Further south there were landing strips at Moulmein, Tavoy, Mergui and Victoria Point.

On their arrival at Mingaladon the pilots of 67 Squadron had settled down to intensive training and to familiarising themselves with the area over which they were to operate. They had little time for either, however, for on 7 December they learned of Japanese landings in Thailand and a state of operational readiness was ordered. Three days later the American Tomahawk squadron arrived at Mingaladon.

The Japanese Army Air Force launched its attack on Rangoon on the morning of 23 December with a force of

60 aircraft. Buffaloes of 67 Squadron and Tomahawks of the Flying Tigers met them in aerial combat.

Ground control was taken by surprise when over the R/T there came not the familiar cry of 'tally-ho' but the excited voice of Sergeant C. V. Bargh, RNZAF: 'Hell. Showers of them. Look Willie, showers of them!' This unconventional announcement signalled the start of a lively engagement in which 13 Japanese aircraft were destroyed, with others listed as probables.

In the meantime, with his aircraft full of holes, Sergeant Bargh dived away from the enemy fighters, flew out to sea and regained height to await the return of the bombers. His windscreen had oiled up but he took off one of his flying boots, wiped the perspex clean with a sock and turned in to attack the bombers as they came away from the target. Joined now by Sergeant E. H. Beable, RNZAF, at 17,000 feet he dived on the enemy formation and succeeded in destroying a bomber and probably a fighter as well. Beable fired a long burst into a bomber which was last seen trailing smoke.

Sergeant G. A. Williams, RNZAF, of No. 67 Squadron, RAF. He shot down a Japanese aircraft in the first air battle over Burma on 23 December 1941.

On that morning Sergeant Bargh and Sergeant G. A. Williams, RNZAF, were flying a routine reconnaissance patrol when they spotted the invading Japanese force. Bargh's excited report gave 67 Squadron adequate warning. Sergeant Gordon Williams' own words describe the first few minutes of the engagement:

> The Japanese had one lone bomber a few miles behind the formation . . . Bargh and I went to the back of the formation and Bargh said: "I'll take him." It had been drilled into us in our training that our primary object was to shoot down the bombers. Bargh by his action had drawn the fighters away from the formation and I saw them jump him. I saw his Buffalo being riddled with bullets.

Sergeant Williams also saw his opportunity and proceeded to carry out copy book attacks on the Japanese bombers, shooting down one and getting bursts of machine gun fire into the petrol tanks of perhaps another six. As the Japanese tanks were not self-sealing it is probable that some of these aircraft did not return to base but Sergeant Williams claimed no more than the one he had seen go down in flames.

Pilot Officer C. McG. Simpson, RNZAF, was airborne for more than two hours intercepting the heavy bombers. He shot down an Army type 97 and got satisfaction from doing so as the Japanese crew were preoccupied in trying to strafe an American pilot who had baled out and was descending by parachute.

Of the 13 enemy aircraft destroyed in this engagement No. 67 Squadron was able to claim a certain six and three probables. Nevertheless, many of the bombers had succeeded in getting through to the target and both Mingaladon airfield and Rangoon itself were heavily attacked. At Mingaladon a bomb had demolished the operations room, two airmen had been killed and two Buffaloes destroyed on the ground, but morale was high and all hands set to work to build another operations centre. It was completed the next day.

At 11 a.m. on 25 December the warning system reported 120 enemy aircraft heading for Rangoon from the direction of Mergui. The Buffaloes and Tomahawks were scrambled immediately. This time the Japanese had dispatched some 80 fighters with their raiding force. Although heavily outnumbered, the defenders met them on the way in but the fighter opposition was intense and only Williams and an RAF pilot got through to the bombers. They shot down one, each claiming a half share. Williams then attacked a fighter, raked it from end to end, and saw it go earthwards out of control. As he was then attacked from behind he was unable to see it hit the ground.

The Tomahawks of the American Volunteer Group had a field day. They claimed 21 aircraft destroyed, giving a total of 27 aircraft destroyed, two probables and two damaged. Thus, in the course of two major raids, the Japanese lost more than 40 aircraft out of about 180, against which had been pitted no more than 25 British and American fighters! No. 67 Squadron lost Sergeants J. MacPherson, E. B. Hewitt and R. P. MacNabb, all RNZAF pilots, and an RAF officer.

Pilots of No. 67 Squadron photographed alongside a Brewster Buffalo, Mingaladon, Burma, 1941. At rear: Sergeant C. V. Bargh, Pilot Officer C. McG. Simpson. In front, standing: Sergeant G. A. Williams, Sergeant E. E. Pedersen, Flight Sergeant E. L. Sadler, Flight Lieutenant A. Bingham-Wallis, RAF, Pilot Officer G. S. Sharp, Sergeant K. A. Rutherford, Sergeant J. MacPherson.

Two RNZAF pilots, Sergeant E. E. Pedersen and Pilot Officer G. S. Sharp, found themselves being attacked by a number of Japanese fighters but fought their way through after shooting down three of the enemy. Sharp's aircraft was badly damaged; but for the armour plating around the cockpit he would probably not have survived the encounter. He force landed at Zayatkwin, a satellite field of Mingaladon, 28 miles distant, with his instruments inoperable, elevators damaged, several cables severed and a bullet hole in his ammunition tray.

Sergeant Beable made the most of his opportunities, shooting down a Zero that was on his leader's tail, possibly destroying another and damaging a third. Sergeant J. G. Finn, RNZAF, was attacked before he could reach the bombers but he was able to fire a burst into the wing root of a Japanese fighter and saw flames leap from its fuel tank.

Sergeant Bargh was caught on the climb but fired an accurate burst into one fighter which began to trail smoke. He then had to dive to shake off a Zero which had pounced on his tail and by the time he had climbed back the enemy aircraft had disappeared and he had to be content with a probable. Sergeant K. A. Rutherford, RNZAF, raked a bomber but was jumped by three Zeros and had to take evasive action so did not witness the result.

The Japanese did not appear inclined to face a repetition of this sort of fighting and Rangoon remained virtually free from attack until 23 January 1942. Thus, in the opening campaigns of the air war in Burma, a small Allied force achieved a very creditable victory which ensured the safe disembarkation of reinforcements at Rangoon. The Supreme Commander, General Sir Archibald Wavell, and the Governor of Burma, Sir Reginald Dorman-Smith, visited Mingaladon, and warmly congratulated the squadron personnel of all ranks on their splendid showing against the enemy.

Because of the importance of defending Rangoon the Air Officer Commanding, Air Vice-Marshal D. F. Stevenson, could not spare any aircraft from the small and already depleted force at his disposal to take the war to the enemy and until reinforcements of Hurricanes and Blenheims began to arrive in January he could give very little support to the Army. No. 67 Squadron carried out an occasional photographic reconnaissance to obtain information on Japanese air concentrations in Thailand and as a result of one such flight the Buffaloes were sent on a strafing attack on Mesoht aerodrome. The attack was a complete surprise. The aerodrome buildings were thoroughly strafed, two aircraft on the ground were set on fire and a large fire started.

A week later, while on patrol over Tavoy, Pilot Officer P. M. Brewer, RNZAF, attacked a single-engined two-seater aircraft. His first burst must have killed the observer as there was no return fire; no evasive action was taken by the enemy aircraft which caught fire and crashed into the hills.

The bonds of comradeship cost Sergeant Jack Finn his life just before the big engagements of late January 1942. Finn and Pilot Officer Paul Brewer had been detailed to make a reconnaissance over Tak and Mesoht, a long patrol which necessitated a refuelling stop. Both pilots landed at Moulmein and refuelled and were just taking off from the airfield when Japanese aircraft attacked. Finn climbed away unscathed but he must have seen that Brewer was in trouble for he turned back and went in pursuit of Brewer's attackers. Both he and Brewer were shot down and killed.

Towards the end of January sorely needed air reinforcements began to arrive. The fighter strength was augmented by 36 Hurricanes which were distributed among Nos 17, 135 and 136 Squadrons; pilots for these squadrons, including several New Zealanders, had already arrived from England and the Middle East. Reinforced by these three fighter squadrons and by No. 113 (Bomber) Squadron, equipped with Mark II Blenheims, the Air Officer Commanding was able to order attacks on occupied airfields in Thailand and so reduce the scale of air attacks on Rangoon.

No. 113 Squadron also inflicted much damage on the enemy at this time. There were six RNZAF aircrew flying with the squadron — Squadron Leader P. D. Smith, a pilot, who had joined the RNZAF in 1938, Flight Sergeant J. Keys, also a pilot, and Sergeants A. M. Dingle, J. H. Beard, E. Brooking and J. B. L. McKenzie, all navigators.

During the last week of January the Japanese made a determined effort to overwhelm the small fighter force at Rangoon. They made repeated daylight attacks on the city and docks, but after losing some 50 bombers and fighters in six days they gave up the attempt and reverted to night raids.

New Zealanders with No. 67 Squadron were again prominent in the air battles. On 23 January Sergeant W. Christiansen, RNZAF, shot down one Japanese fighter and Pilot Officer A. A. Cooper, RNZAF, damaged another. The next day Cooper, Bargh, Christiansen and Flight Sergeant E. L. Sadler were patrolling over Mingaladon when they sighted a formation of enemy bombers to the south-east. They dived to intercept and in the ensuing action shot down every one of them.

On the ground, however, the situation was rapidly deteriorating. The retreating British Army sustained a severe reverse at the Sittang Bridge, and the way was open to Rangoon. The Air Force was forced back to Magwe where the warning system — so vital to aerial defence — was totally inadequate to cope with the possibility of surprise strafing and bombing attacks. As more airfields fell into the enemy's hands so the weight of attacks increased and the situation of the defending fighters became ever more precarious.

As the fall of Rangoon became imminent the remnants of the Allied fighter force — four battle-worn Buffaloes, four American Tomahawks and some 20 Hurricanes — were ordered northwards. The Buffaloes flew to Dum Dum, Calcutta's principal civilian airport, with a stop at Chittagong to refuel. Sharp, Bargh, Rutherford and an RAF Buffalo pilot flew their aircraft out to India, while the remainder of 67 Squadron travelled by transport aircraft. A couple of days after arriving in Calcutta four of the 67 Squadron pilots were sent to Akyab, this time flying Mohawks, but their five-day period of readiness there was uneventful and they were recalled before any large-scale Japanese raids occurred. One of the squadron survivors, Pilot Officer Sharp, was destined to return to Burma three years later in command of a fighter squadron of the Royal Indian Air Force and, in the triumphant liberation of the country from Japanese domination, win the DSO.

Sergeant J. G. Finn, RNZAF, of No. 67 Squadron, who was shot down going to the aid of another New Zealand pilot in Burma, is second from left in this group of New Zealand servicemen being entertained at lunch by a Netherlands East Indies women's organisation in Batavia in 1941 some months before the Japanese invasion.

CHAPTER EIGHT

The New Zealand Squadrons in the Royal Air Force

As the vast network of the Empire Air Training Scheme poured forth its graduates, New Zealand airmen began arriving in Britain in large numbers, but the RNZAF was reluctant to establish and maintain its own squadrons in the Northern Hemisphere. As a separate service it was only three years old by early 1940. It was still being led by an RAF officer on loan (Group Captain Saunders) and its resources were still largely devoted to training. To have created its own squadrons in Britain would have posed weighty administrative and operational problems for the RNZAF. The Australian authorities held similar views, although Canada was insistent that its nationals should serve in RCAF squadrons.

Under Article 15 of the Empire Air Training Plan, the United Kingdom undertook that 'the pupils of Canada, Australia and New Zealand shall, after their training, be identified with their respective Dominions either by organising Dominion units or formations, or in some other way, such methods to be agreed upon with the respective governments . . .'

In the discussions initiated by the British Government towards this end it was agreed that six squadrons should be formed within the RAF, all to be identified with the Dominion and designated 'New Zealand Squadrons'. This was in keeping with the tradition already established by No. 75 (New Zealand) Bomber Squadron. Of the six new units (485 to 490 inclusive), three were formed within RAF Fighter Command, two in Coastal Command and one in Bomber Command. The aircrew components generally comprised members of the RNZAF, New Zealanders serving in the RAF, a leavening of United Kingdom personnel and some representatives of other Allied countries. Ground crew were predominantly RAF, few of them New Zealanders. (In the brief descriptions which follow, the men named were New Zealanders unless their nationalities are specifically stated.)

No. 485 Squadron

No. 485 (New Zealand) Squadron, often referred to simply as 'the New Zealand Spitfire Squadron', was formed at Driffield in Yorkshire on 1 March 1941. Squadron Leader M. W. B. Knight, an experienced flying instructor when war broke out, was given command. He had two knowledgeable flight commanders in Flight Lieutenants F. N. Brinsden and J. C. Martin, both of whom had joined the RAF in 1937. Knight also had two experienced pilots in Pilot Officer E. P. Wells who had fought in the Battle of Britain and Sergeant W. V. Crawford-Compton who had started his Air Force career as an aircraftman.

The combat experience of such men was of incalculable value in moulding the new squadron which had in its ranks many pilots who had recently graduated from the RNZAF's No. 2 Flying Training School at Woodbourne and whose flying experience was limited to Vickers Vincents and Vildebeests and the short introduction to more modern aircraft types afforded by a period at OTUs in Britain.

No. 485 consisted almost entirely of New Zealanders and it was this fact, coupled with its record of courage in action, that led to New Zealanders' intense interest in the squadron, an interest which triggered a campaign to raise funds for the purchase of Spitfires. Begun in Southland, the scheme quickly gained impetus throughout the Dominion and £126,000 was raised. The campaign was organised on a provincial basis and when the squadron was re-equipped with Spitfire VBs, these aircraft were named after New Zealand provinces.

There were New Zealanders in the ground staff, which was largely made up of RAF tradesmen, and some stayed with the squadron for several years while three at least subsequently remustered to aircrew and trained as pilots.

On 12 April the squadron became operational but its first tasks, convoy escort patrols over the North Sea, were not exciting. Squadron Leader Knight scored the first victory when on 2 June he and three other pilots intercepted a section of Junkers 88s attacking a convoy off the Yorkshire coast. After the initial shots had been exchanged the enemy aircraft escaped into cloud but in a subsequent hide-and-seek engagement Knight scored hits which eventually brought one aircraft down into the sea, near the convoy. One of the escort vessels subsequently confirmed that the aircraft had been destroyed.

Under Knight, 485 Squadron took part in shipping escort patrols and offensive sweeps over northern France. Later it gave close escort to daylight bombers. At the end of 1941 the squadron's scoreboard disclosed 18 confirmed victories, six probables and 12 enemy aircraft damaged.

Wells, who had three Messerschmitts to his credit and several probables when he joined 485 Squadron, succeeded Knight as squadron commander in November 1941 when the squadron was newly established at Kenley in Surrey. Wells was a champion rifle shot and this, together with his ability to sight an enemy aircraft before most others, gave him his nickname of 'Hawkeye'.

A Spitfire IX of No. 485 (New Zealand) Squadron RAF, providing fighter cover for returning American bombers.

A deadly Typhoon attack on Channel shipping by aircraft of No. 486 (New Zealand) Squadron.

RNZAF Kittyhawks being serviced by No. 4 SU at Ondonga, New Georgia.

Over burning Singapore, Brewster Buffaloes of No. 488 (New Zealand) Squadron about to engage the enemy.

Pilots of No. 485 Squadron at Kenley after a successful attack on German battleships. From left: H. N. Sweetman, D. T. Clouston, B. E. Gibbs, R. J. C. Grant, M. M. Shand, E. P. Wells, I. J. McNeil, W. V. Crawford-Compton and J. M. Checketts. *Topical Press Agency Ltd, London*

Sergeant Crawford-Compton, who was destined to reach the rank of air vice-marshal in the RAF, proved himself as a member of 485 Squadron. By May 1942 he was a flight lieutenant and had shot down five enemy aircraft and shared in the destruction of another. A crash landing in which he broke a wrist put him out of the running for command of the squadron but this was but a slight check to his progress. (By May 1945 he was a wing commander and one of New Zealand's most decorated combat pilots with the DSO and Bar, the DFC and Bar, the Silver Star (US), the Croix de Guerre with Palm and the Legion of Honour. He had 21½ enemy aircraft destroyed to his credit. He elected to make his postwar career in the RAF and by 1968, as an air vice-marshal, he had the CB and CBE added to his decorations. One of his citations described him as having 'a rare zest for battle'.)

In the summer of 1943, No. 485 Squadron, re-equipped with Spitfire IXBs, provided high escort cover for some of the big daylight bombing raids, several of which included USAAF squadrons.

After leading 485 with distinction Wells was promoted to command the wing in which the squadron flew. He was succeeded as squadron commander in May 1942 by Squadron Leader R. J. C. Grant, an original member of the squadron who had also experienced rapid promotion. Under Grant, 485 Squadron further raised its score. In January 1943 the squadron moved to the No. 11 Group station of Westhampnett (a satellite of Tangmere), which was commanded by a distinguished New Zealander serving in the RAF, Group Captain H. D. McGregor, DSO, a Battle of Britain veteran. (McGregor also had No. 486 Squadron under his command at this time.)

In February 1943, 485 Squadron was engaged over Abbeville in a fierce conflict with 50 Messerschmitts and Focke Wulfs. To his great dismay Grant saw a Focke Wulf attack his brother, Pilot Officer Ian Grant, and, despite his best efforts, he had the tragic experience of seeing his brother shot down. He also had the satisfaction of shooting down the German aircraft responsible almost at once. This was Grant's last sortie with 485. He had then completed 150 operations and for a period he went to Canada where he lectured to Air Force training units. Back in the United Kingdom he was promoted to wing commander and returned to operations when he was given command of a wing of Mustang fighters. He was killed on operations in February 1944.

When Grant went to Canada he was succeeded as commanding officer of 485 Squadron by Squadron Leader R. W. Baker, who had been one of his flight commanders, but Baker had been with the squadron for almost two years and was more than due for a rest. He remained with the squadron until June 1943, rounding off a period of commendable service. His successor was Squadron Leader J. M. Checketts, an RNZAF officer who had earlier served with 485 Squadron and more recently had been a flight commander of No. 611 Squadron at Biggin Hill. At this famous RAF station, the wing was led by Alan Deere, who had recommended Checketts to lead the New Zealand Spitfire squadron when it was re-equipped with new Spitfires. Although Checketts was then 32 and therefore older than most fighter pilots, he was known as a 'natural' fighter pilot and a good leader in the air. By the time he was himself shot down in September he had 11 German aircraft to his credit.

485 Squadron joined Deere's wing in June 1943, and during July and August it destroyed more enemy aircraft than any other squadron in No. 11 Group. Checketts was shot down the following month by a formation of Focke Wulfs near Lille. He was seen to land by a French boy who cycled to the spot, ignoring several hundreds of German soldiers who were also searching for the downed airman. Checketts was badly burned about the face and hands. The boy reached the pilot first and took him to his home where he was helped to escape. Checketts closed his wartime operational career when he led a Spitfire wing into fierce action against German fighters over Arnhem in September 1944.

There was a time in its colourful career when one

section in 485 Squadron was made up of New Zealand farmers — Flight Lieutenant J. G. Pattison, DFC, Reg Grant and Flight Lieutenant Baker. Pattison, one of the 'Few' who had fought in the Battle of Britain, was promoted to squadron leader and became 485's seventh commanding officer in September 1944. At the time of his appointment he had 300 operational hours to his credit.

On 26 August 1945 No. 485 Squadron was disbanded at the Luftwaffe airfield of Fassberg. During its operational career the squadron flew 10,717 sorties, destroying 63 enemy aircraft in combat, probably destroying another 25 and damaging 32. During the brief period that the squadron was employed against ground targets some 70 motor vehicles and five railway engines were destroyed.

Its links with the RNZAF through the Empire Air Training Scheme and the RNZAF members who flew with and led the squadron were cemented when, in 1967, the Squadron Association presented its official badge to the RNZAF Strike Wing at Ohakea.

'Achtung! Spitfire.'

New Zealand pilots of No. 485 Squadron being introduced to King George VI by Air Marshal Sir Sholto Douglas, AOC RAF Fighter Command. From left: S. F. Browne, B. E. Gibbs, J. R. C. Kilian and E. D. Mackie. *Associated Press*

Flight commanders of No. 485 Squadron, Flight Lieutenants M. R. D. Hume and M. G. Barnett, with the New Zealand High Commissioner in Great Britain, Mr W. J. (Bill) Jordan (centre), Squadron Leader J. M. Checketts (second from right) and Wing Commander A. C. Deere, RAF who was Wing Commander Flying at Biggin Hill. *Sport and General*

New Zealand ground crew members of No. 485 Squadron. From left: Sergeant F. M. B. George, Corporal T. G. Smith, Leading Aircraftmen T. H. MacGibbon, E. S. Bongard, Aircraftman G. B. Neville, Leading Aircraftman G. E. Erridge, Corporal N. Murray, Aircraftman A. W. Martin and Corporal G. H. French. *Sport and General*

Armourers of No. 485 Squadron cleaning and reloading machine guns. *Sport and General*

No. 486 Squadron

No. 486 Squadron was formed at Kirton-in-Lyndsey, Lincolnshire, on 7 March 1942, initially with Hawker Hurricane aircraft, but it re-equipped with Typhoons in July that year and with Tempests in 1944. Like 485 Squadron this unit won many honours, gaining special distinction for the shooting down of the dreaded flying bombs launched by Germany against southern England in mid 1944.

No. 486 Squadron was first led by an Englishman, Squadron Leader C. L. C. Roberts, who had fought in the Battle of France. His two flight commanders were New Zealanders, however — Flight Lieutenants J. G. Clouston and H. N. Sweetman. Sweetman had seen extensive service with 485 Squadron and was combat experienced.

In August the squadron was re-equipped with Typhoon 1B aircraft. The unit, which had been transferred to Wittering soon after its formation, moved to North Weald in September and then to West Malling in October. There the squadron operated in the day fighter role, flying on coastal patrols and escorting light bombers which were attacking coastal targets in German-occupied territory.

On 29 October the squadron with its Typhoons moved to Tangmere in Sussex, its third move in four weeks. The Typhoon was powered by a huge 24-cylinder Napier Sabre liquid-cooled engine and its gigantic three-bladed propeller gave it a somewhat pugnacious air. Whereas the Hurricane was a beautiful aeroplane to fly, the newer and faster steed from the same stable was not so easy; it had to be mastered.

Mastering the Typhoon was his first challenge when command of 486 Squadron passed to Squadron Leader D. J. Scott, DFC and Bar, RNZAF, on 1 April 1943. After two tours of operations in Hurricanes Scott had been 'rested' with a temporary staff appointment at Bentley Priory, Headquarters of RAF Fighter Command, and he assumed command of 486, pleased to be back on operations. He was then 24 years of age and had risen from the rank of sergeant to squadron leader in a mere nine months, an indication of both the heavy casualties which Fighter Command had suffered and his own remarkable abilities.

When Scott arrived at Tangmere he found the station was commanded by a New Zealander, Group Captain H. D. McGregor, RAF, while his group commander was another RAF officer who had ties with New Zealand —

Wing Commander D. J. Scott with his personalised Typhoon, September 1943. *Air Ministry*

Air Vice-Marshal Sir Hugh Saunders who had served as the RNZAF's Chief of Air Staff from 1939 to 1941.

The frequency of operational patrols across the English Channel was increased, an offensive rather than a defensive role was assumed, and under Scott the squadron in the first two weeks accounted for two German aircraft destroyed, several others damaged, two minesweepers set on fire, four ships damaged and an oil tank set on fire. As commanding officer, Scott continued to set a pattern of dynamic leadership and aggressiveness.

Seven German E-boats were either destroyed or damaged on 3 August and a coastal vessel was sunk off Cherbourg 12 days later, the squadron meeting heavy anti-aircraft fire during these operations. At the end of August Squadron Leader Scott was awarded the DSO and promoted to wing commander. By November 1944, at the age of 26, Scott enjoyed the distinction of being the youngest group captain in the RNZAF, having gained this rank after two years of commissioned service. His decorations then included the DSO, OBE, DFC and Bar. His OBE was awarded in 1944 for gallantry in rescuing a pilot from a Spitfire which had caught fire on landing. Later he received recognition for his leadership during the period that his wing of rocket-firing Typhoons operated from the Caen bridgehead and further honours included the Croix de Guerre and Palm from both France and Belgium and being made a Commander of the Order of Orange Nassau by Holland.

An original RNZAF member of 486 Squadron with an especial claim to fame was Squadron Leader A. E. Umbers who, at the time of his death, was credited with the destruction of 28 flying bombs. Umbers, who had completed a tour of duty with 486 Squadron as a junior officer, took command of the squadron in his second tour in December 1944. The squadron was then flying Britain's latest and fastest fighter, the Tempest, and was serving with the Second Tactical Air Force as part of the renowned 122 Wing, commanded by a New Zealander, Group Captain P. G. Jameson, RAF, and was stationed at Volkel in Holland. About this time ground staff and pilots alike were subjected to attacks on the airfield by Me262 jet fighters carrying anti-personnel bombs. No. 486 retaliated by destroying several 262s in the air.

Throughout the Ardennes campaign the squadron supplied constant patrols over the battle area and suffered few losses. On 14 February 1945, Umbers' aircraft was hit by flak and he was killed when the aircraft went down in flames. No. 486 Squadron suffered several casualties about this time and Fighter Command was unable to keep it supplied with sufficient New Zealanders trained to operational standard on the Tempest. It was necessary, therefore, to post in a few Australians and, ultimately, some English pilots to help man the unit until it was disbanded at Dunsfold, in England, on 20 September 1945, after being stationed in Copenhagen and then at Lubeck in Germany in the last months of the war.

By the end of the war No. 486 Squadron had positively destroyed 81½ enemy aircraft in the air (and three on the ground), probably destroyed another five, damaged a further 22, shot down 223 flying bombs and destroyed or damaged much enemy shipping. It was a magnificent record for the New Zealand unit.

No. 487 Squadron

No. 487 (New Zealand) Squadron was formed at Feltwell in August 1942 under Wing Commander F. C. Seavill, a New Zealander in the RAF, and was equipped with Ventura light bombers. Seavill, who had been serving since 1930, did not command it for long; he was killed on operations a few months later. Feltwell, the station where 75 Squadron had begun operations against Germany in 1940, was then commanded by a New Zealander, Group Captain R. L. Kippenberger, RAF. From the beginning 487 Squadron had a strong Empire representation. Most of the ground and administrative staff were drawn from the RAF but there were several New Zealanders among

Squadron Leader L. H. Trent, VC, DFC. He learned to fly in the RNZAF and joined the RAF in 1938.

Ventura bombers of No. 487 (New Zealand) Squadron attack the Philips factory at Eindhoven on 6 December 1942.

them and more were to follow. The first flight commanders were Squadron Leader L. H. Trent, who held an RAF commission, and Squadron Leader A. B. Wheeler, also a New Zealander, who had the unusual distinction of belonging to the RCAF in which he had enlisted on the outbreak of war.

Over three years of operations, the squadron built up a reputation under the leadership of men such as Wing Commander G. J. Grindell, AFC, RAF, who succeeded Seavill; Wing Commander A. G. Wilson, DFC, an Englishman who saw the squadron convert to Mosquito day-bombers; and Wing Commander I. S. Smith, DFC and Bar, a survivor of the Battle of Britain. No. 487 knew many great moments but will probably be best remembered for Trent's daring low-level attack on a power station at Amsterdam in 1943. This action earned this man from Nelson the Victoria Cross for courage and fine leadership against massive odds.

The Venturas took off from their base at Methwold in Norfolk on 3 May in the late afternoon. They flew in two formations. At the briefing the crews had been told that they would take part in a series of attacks designed to help the Dutch Resistance movement and encourage strikes against German authority. No. 487 Squadron was to bomb the power station and at the same time create a diversion while a formation of Bostons struck against the power station at Ijmuiden.

'Your target is well defended,' the briefing officer had warned. 'It is most important, however, that this attack be pressed home regardless of opposition.' Before taking off Trent was heard to say that he was going to reach that target no matter what happened. All went well until the Dutch coast came in sight. Two covering Spitfire squadrons upset the plan when they arrived over the Dutch coast too early and at too great a height. They thus alerted the German radar and enemy fighters were able to assemble in strength. They were waiting, at altitude, to confront the main British attacking force.

The 11 Venturas, with their supporting fighters, were pounced on at 12,000 feet. The aerial ambush was staged by more than 70 German fighters. The FW190s dived on the Spitfires and Me109s engaged the bombers. Quickly assessing the situation — and the odds — Trent ordered his crews to close into a single formation for protection and to proceed 'all-out' to the target, now visible just ahead. In the bitter and confused fighting which followed the Venturas were picked off one by one, until only five remained to commence the bombing run. Trent shot down one assailant but the Germans were now queueing up above and taking turns at diving on the bombers. One after another the Venturas were swiftly shot down until finally only Trent was left. He nevertheless flew on to bomb the target. His bombs overshot but were near enough to the target to cause blast damage. As he turned towards the coast to run the gauntlet of the German fighters once more, his aircraft was hit and went down. The bomber broke up in the air and Trent and his navigator were thrown clear. They survived and became prisoners of war.

Of the 11 Venturas which set off that spring afternoon only one returned to base — an aircraft which had been severely damaged in the first engagement. Of the 10 aircraft destroyed, few of their crew members survived. Almost all the casualties were members of the RNZAF. The citation to Trent's VC concluded with the words: 'On this, his twenty-fourth sortie, Squadron Leader Trent showed outstanding leadership. Such was the trust placed in this gallant officer that the other pilots followed him unwaveringly. His cool unflinching courage and devotion to duty, in the face of overwhelming odds, rank with the finest examples of these virtues.'

The Bristol Beaufighter. This aircraft served both No. 488 (New Zealand) Squadron in the night-fighter role and No. 489 (New Zealand) Squadron in RAF Coastal Command. *Air Ministry*

No. 488 Squadron

The first 488 Squadron had been disbanded in New Zealand in March 1942 after its withdrawal from the Far East. The second 488 Squadron retained its New Zealand ties when it was formed at Church Fenton, Yorkshire, on 25 June 1942, and began training on Beaufighter aircraft in the night-fighter role under Wing Commander R. M. Trousdale. Trousdale, although only 21, had already destroyed eight enemy aircraft, five of them at night, and had won both the DFC and Bar. He had joined the RAF at the age of 18 and was to transfer to the RNZAF in the final year of the war. He had two first-class flight commanders in Squadron Leaders J. R. Gard'ner and P. W. Rabone. Gard'ner had been on night-fighter operations, flying first Blenheims and then Defiants, had force landed in the Channel and had spent three months in hospital before returning to active service. He came to 488 Squadron with Trousdale from the same Canadian Beaufighter squadron. Rabone had also seen a great deal of action. He had flown Fairey Battles in France and had fought in the Battle of Britain.

On 1 September the squadron moved to Ayr in Scotland, and a detachment was sent to Drem, on the east coast, for the interception of enemy raiders. There were 21 New Zealand aircrew in the squadron, almost all members of the RNZAF.

When Mosquito night-fighters replaced the Beaufighters in September 1943, No. 488 Squadron was based in Essex where the station commander was Wing Commander R. F. Aitken, a New Zealander serving in the RAF. The squadron had established a fine reputation for its successful raids on the German railway system* — the Beaufighter with its impressive firepower of four cannon and six machine guns, was very suitable for such missions

* The most successful 'train busting' team in the squadron was Pilot Officer G. F. Reed and his English navigator, Flight Sergeant R. Bricker, who claimed 13 engines destroyed or damaged.

Servicing Beaufighters of No. 488 Squadron.

The Mosquito night-fighter which replaced the Beaufighter late in 1943.

— but now the emphasis was on intruder missions over enemy territory and night defence of its own area. In the transitional period the New Zealanders received much help from No. 605 Squadron, RAF, which shared the same airfield and operated in the same role. In their intruder patrols, 488 Squadron crews ranged as far afield as the Baltic on special missions by day as well as by night.

When the Normandy invasion began the squadron had shot down 20 enemy aircraft. During that historic month of June 1944, another nine were added to the score and by the end of the year the total had increased to 62. Many of these actions were by night.

Several Englishmen commanded the squadron after Trousdale but in 1944 and 1945 it was led by Wing Commander R. G. Watts, a very popular commanding officer, who had joined the RNZAF in December 1940. He was in command when it was based at Amiens-Glisy in France from November 1944 to April 1945 and at Gilze Rijen, Holland, where the squadron concluded its operational career. Many decorations for squadron crew members were advised before the unit disbanded at the end of April. The squadron flew its final patrols of the war on 25 April 1945 when the last aircraft to land reported the destruction of an FW189 — a fitting end to this squadron's highly successful career.

One of the leading night-fighter units of the Second Tactical Air Force, No. 488 Squadron on its disbandment could claim 67 enemy aircraft destroyed, three probably destroyed and 11 damaged, besides 40 enemy trains and many road vehicles, barges and ground installations destroyed.

No.489 Squadron

The success of 489 (New Zealand) Squadron, a torpedo-bomber unit of RAF Coastal Command, was most apparent in the volume of enemy shipping which it sent to the bottom of the sea. Formed at Leuchars in Scotland in August 1941 with Beauforts and later with Blenheims, it did not become operational until April-May 1942 when it was equipped with 20 Hampden bombers and had 68 New Zealanders on its strength. Later it flew Beaufighters.

Wing Commander J. S. Dinsdale, DFC, was the first New Zealander to command the squadron after earlier serving with it as a pilot. The third commanding officer, he was awarded the DSO shortly before he handed over command to Wing Commander L. A. Robertson, who led the squadron with distinction in 1944–45.

From the middle of 1944 the squadron operated as part of an Anzac Beaufighter Wing which included Australian

Hampden bombers of No. 489 (New Zealand) Squadron flew almost to the limit of their range in anti-shipping strikes off the Norwegian coast.

and British squadrons and in the last year of the war the New Zealanders flew 1250 sorties, mainly in patrols and attacks over the North Sea and along the Norwegian coast.

In October 1944 the Anzac Wing moved north and for the rest of the war it operated over Norwegian waters from the airfield at Dallachy, some 30 miles east of Inverness. Here, on the north-east coast of Scotland, the airfield was exposed to the full force of the frequent gales which swept in from the North Sea, so that aircraft had to be picketed down or else protected by small hangars or earthen mounds. When the winter storms were at their height it was difficult to move about the airfield and the wide dispersal of buildings and Nissen huts where crews slept did not help matters. 'We shall not easily forget the last winter of the war,' wrote one pilot. 'Taking off from

the icebound or snow-covered field we sometimes flew 400 miles across the North Sea to find the Norwegian coast shrouded in mist or low cloud. Flying among the islands and into the fiords in search of ships hiding there was rather hazardous, since some of the passages were so narrow that there was little room in which to manoeuvre.'

Despite the difficulties of both climate and terrain, however, operations were maintained at a high level, with crews of 489 Squadron taking part in many large strikes and also flying rover patrols by day and night in search of targets.

In December 1944 Squadron Leader D. H. Hammond, who was a flight commander of 489 Squadron, received an immediate award of the DSO after a most successful attack on Den Helder harbour which inflicted widespread damage on enemy docks and shipping. Hammond, who had already been decorated with the DFC and Bar, was subsequently promoted to the command of the squadron as a wing commander, and he led it through the final stages of the war.

From 489's formation until April 1944, its crews were responsible for the destruction of 11 ships totalling 38,700 tons and the damaging of a further 13 ships totalling some 40,000 tons. From then on until the end of the war in Europe 489 Squadron shared, as part of the Anzac Strike Wing, in the sinking of a further 19 ships of 67,000 tons and 12 escort vessels. A further 18 cargo ships, together with 49 escorts of various types, were damaged. The cost of these successes was not light; the squadron lost 31 aircraft during its operational career, few of the crews surviving.

New Zealanders with the squadron won two DSOs, 19 DFCs, one Bar to the DFC, one DFM and one Conspicuous Gallantry Medal.

No. 490 Squadron

No. 490 (New Zealand) Squadron was a Coastal Command unit which was formed at Jui, near Freetown (West Africa) in March–April 1943, with an advanced base at Fisherman's Lake in Liberia. First in Catalina flying-boats, and later in Sunderlands, 490 Squadron flew anti-submarine patrols, convoy escorts and air-sea rescue missions, soon achieving an enviable reputation for operational efficiency and aircraft serviceability which it was to maintain throughout its sojourn in West Africa to the end of the war, despite the exacting and monotonous nature of much of its later work.

Group Captain B. S. Nicholl who, as a wing commander, commanded No. 490 (New Zealand) Squadron in West Africa in 1943–44.

Thomas Francis (Frank) Gill, DSO, a postwar photograph taken when he was a group captain.

Wing Commander D. W. Baird, the first commanding officer, had joined the RNZAF in 1938 after service in the RAF and had commanded the RNZAF in Fiji before going to West Africa. He was followed by Wing Commander B. S. Nicholl who had been a journalist in civilian life and had also commanded the RNZAF in Fiji earlier in the war. The third and last commanding officer was Wing Commander T. F. Gill, DSO, who assumed command in Sep-

Catalina flying-boats of No. 490 (New Zealand) Squadron at Jui, West Africa, in early 1943.

tember 1944, bringing with him a distinguished record as a bomber pilot, including service with 75 Squadron.

Because the West African waters were so well patrolled in the later stages of the war, few U-boats ventured to operate there and those on passage to the Indian Ocean usually passed well out of range. Crews therefore had to face the long periods of boredom that lack of action produces, besides a tropical climate with its sudden, violent and treacherous storms. When one Sunderland was forced down 200 miles out to sea, the pilot, Flying Officer M. E. McGreal, made a fine landing but the aircraft broke up quickly in the heavy swell. Dinghies were launched but two of the crew were unable to get clear. Warrant Officer R. A. P. Opie, the wireless operator, was last seen at his post trying to transmit distress signals. In the water Sergeant D. C. Jones, an RAF crew member, supported an injured comrade until McGreal swam across to inflate the second dinghy and retrieve the paddles. Without food or water, the survivors drifted until the next evening when they were sighted and rescued.

No. 490's principal task was the escorting of convoys and this duty was performed with creditable efficiency. Early in March 1945 when the long immunity enjoyed by Allied shipping in West African waters was broken suddenly with the torpedoing of a ship 500 miles off Walvis Bay, Gill's squadron headed the list of units in the area with the largest total of hours flown in the hunt for the German U-boat.

Wing Commander Nicholl, in a report on the squadron's work during his term as CO, aptly summed up the role of the anti-submarine warfare unit — 'This game is rather like cricket. The runs saved in the field count just as much towards winning the game as the runs made off the bat.'

Wing Commander H. L. Thompson, in his official history, *New Zealanders with the Royal Air Force*, summed up the squadron's contribution in these words: 'No. 490's main achievement, like that of so many Coastal Command squadrons, can best be measured in terms of Allied ships saved from U-boat attack and the safe arrival of troops and supplies at their various destinations.'

No. 490 Squadron flew its last patrols in the closing stages of the European war in defence of shipping off the coast of West Africa. The unit was disbanded in June 1945. Throughout its operational life the squadron flew 463 sorties totalling 4853 flying hours.

RNZAF Aircrew in Other RAF Squadrons

While men like Scott, Umbers, Grant and Checketts were distinguishing themselves as members of the 'New Zealand Squadrons' other RNZAF aircrew were making a mark in RAF squadrons not so well known in New Zealand. (The wartime information services ensured that the '480 series' units and the New Zealanders serving with them were widely and regularly publicised.)

Squadron Leader J. R. C. Kilian, a 485 Squadron pilot who had joined the RNZAF in 1937, commanded No. 122 Squadron in 1942 and No. 504 Squadron the following year and was decorated with the French Croix de Guerre. Squadron Leader M. G. Barnett, after serving with 485 Squadron as a junior officer went on to command No. 234 Squadron in 1943, 501 Squadron in 1943–44 and 274 Squadron later in 1944 and was awarded the DFC. Wing Commander W. H. Stratton, DFC and Bar, commanded No. 134 Squadron, an RAF fighter squadron, in the Middle East, Burma and India. Wing Commander G. D. Size, who joined the RNZAF just after the outbreak of war, commanded No. 248 Squadron and RAF Station Mount Farm and won the DSO and Bar and the DFC and Bar. He transferred to the RAF after the war. Wing Commander J. F. Barron, who joined the RNZAF in July 1940, went from sergeant to wing commander in three years and was commanding No. 7 Squadron, RAF, when he was killed in 1944 on his 78th mission with Bomber Command and his 32nd with the Pathfinder Force. Barron had won the DSO and Bar, the DFC and DFM when his Lancaster was shot down in pre-D Day operations.

One name that cannot be omitted is that of Squadron Leader R. B. Hesselyn, MBE, DFC, DFM and Bar, who joined the RNZAF late in 1940. In operations from Malta, flying Spitfires, he quickly made a name for himself and was credited with the destruction of 12 enemy aircraft during early 1942. The following year he shot down a further seven aircraft while he was operating out of Hornchurch. At the end of September 1943 Hesselyn was the highest scoring fighter pilot in the RNZAF with 18½ confirmed victories. In 1946, on his return to England, Hesselyn was made an MBE for 'distinguished services while a prisoner of war in Germany'. (Pilot Officer James Barnes, a 75 Squadron navigator who was shot down in 1942 and spent the rest of the war in POW camps, was similarly honoured that same year.)

The story of the RNZAF in the European theatre would not be complete without mention of Flying Officer L. A. Trigg, DFC, who, with Ward and Trent, joined the ranks of Victoria Cross winners in the Second World War. Trigg, who was born in Houhora, joined the RNZAF in 1941. On his first operational sortie in a Liberator Bomber of Coastal Command (he was then serving with No. 200 Squadron) he took off from Rufisque near Dakar shortly after dawn on an August day in 1943 to patrol an area where U-boats had been reported. Trigg spotted a surfaced U-boat after four hours of patrolling and attacked at once. The German did not submerge but instead engaged the Liberator with its anti-aircraft guns and scored repeated hits on the bomber, which caught fire on its approach run. Trigg continued on his course — he could have broken off the engagement and landed in the sea — and straddled the U-boat with depth charges dropped with fine accuracy. When he was attacking at less than 50 feet, heavy anti-aircraft fire entered through his open bomb doors. The U-boat sank a few minutes later. Almost at the same time the flaming Liberator crashed into the sea and the crew perished. (Survivors of the German U-boat included the commander, who gave an account of the engagement and expressed admiration of Trigg's courage.)

Corporal C. B. G. Knight earned the distinction of being the first member of the RNZAF to win a decoration in the Second World War. The event that led to the award predated his joining No. 75 Squadron by a few months. Squadron Leader A. McKee, who had left New Zealand to join the RAF in 1926, was leading a formation of Wellington bombers on a search in the Heligoland Bight on 14 December 1939. Flying through storms, with heavy cloud down to within a few feet of the sea, the Wellingtons located a large enemy naval force. As they approached their target the bombers met heavy anti-aircraft fire and then a formation of 50 German fighters. Three Wellingtons were shot down in flames and a fourth was damaged and crashed when attempting to land back at base. Despite the weather and the opposition, McKee's formation identified every unit of the German Fleet and made a reconnaissance of the whole area. Knight, who was chief wireless operator for the entire 12-aircraft formation, obtained useful bearings from German stations and passed messages to base, including sighting reports of the enemy ships. Not only did he do this while under fire but he also manned one of the guns in between sending messages. It was for these actions that he was awarded the DFM.

There were many, many others, too numerous to mention, whose daring, skill and courage, combined with discipline, contributed to the reputation of New Zealand airmen. That great Welshman, Wing Commander Ira Jones, DSO, MC, DFC, MM, RAF, said in his book *Tiger Squadron*, 'When I think of all the Commonwealth pilots I have met in two world wars I know that for my ideal squadron I would choose first New Zealanders, then South Africans, then Canadians and then Australians.'

Wing Commander Colin F. Gray, New Zealand's highest-scoring wartime fighter pilot, had 27½ enemy aircraft destroyed to his credit, besides eight probables. He was decorated with the DFC and two Bars and the DSO. *Sport and General*

As many as 10,950 members of the RNZAF served with the RAF in the Second World War, and although seven RAF squadrons were identified with New Zealand the majority of these airmen were scattered throughout units in the United Kingdom, and in various theatres of war. They won many honours but they also suffered heavy casualties. No fewer than 3285 were killed in action, another 548 became prisoners of war and a large number suffered wounds. In all, their service was such that the people of New Zealand could read with pride Britain's Air Council message at the end of hostilities which paid tribute to the illustrious part New Zealand airmen had played and the honour they had brought to this country by their gallant service in all theatres of war.

Flying Officer L. A. Trigg, VC, DFC, RNZAF.

CHAPTER NINE

Preparations for a Pacific War

Two far removed from Great Britain to expect immediate aid from that quarter, New Zealand, in the dark days that followed Pearl Harbour, found herself confronted with the threat of invasion as the Japanese war machine struck swiftly southward.

On 2 January 1942 the Japanese Army entered Manila. Two days later Rabaul was raided by aircraft from the formidable Japanese naval base at Truk. With hardly a check, the enemy advanced on all fronts. By June the Japanese were firmly established on a line stretching from the Andaman Islands in the Indian Ocean through the Netherlands East Indies to the Central Solomon Islands. Darwin had been attacked by Japanese warplanes, to the alarm of the Australian people and the consternation of New Zealanders who felt suddenly the war's closeness to their own shores. New Guinea was threatened.

The ease and rapidity with which Japan had thrust southward brought about a complete revision of the thinking on New Zealand's defence requirements. Instead of facing the likelihood of danger from occasional German sea raiders, New Zealand now had to contemplate the possibility of enemy invasion. From its role as primarily a training organisation, the RNZAF had additionally and quickly to develop an operational air force capable of meeting the enemy in combat.

A week after Japan entered the war the British Prime Minister, Mr Winston Churchill, had flown to Washington to confer with the American authorities on the problems of the Allied Command. The United States, in conjunction with Australia, New Zealand and the Netherlands East Indies, was made responsible for the Pacific area. The strategic direction of the war as a whole, excluding operations in Russia, was placed in the hands of the Combined Chiefs of Staff Committee in Washington which comprised the three American Chiefs of Staff and a British Joint Staff Mission. A unified command was created in the South-West Pacific under General Sir Archibald Wavell with an American deputy commander who was in charge of the naval forces in the area.

Then, early in March, after Java had fallen to the enemy, the United States proposed that the whole of the Pacific to a line west of Singapore should be the responsibility of America and the American Chiefs of Staff. Finally, the Pacific region was split up into the South-West Pacific under General Douglas MacArthur and the Pacific Ocean area under Admiral Chester W. Nimitz. The South-West Pacific area comprised the Philippines, Borneo, the Netherlands East Indies east of Sumatra, New Guinea, New Britain, the Northern Solomons and Australia. The Pacific Ocean area, which was subdivided into Northern, Central and South Pacific areas, comprised the rest of the Pacific to the west coast of America so that New Zealand, which was in the South Pacific, was placed in a different command from Australia. The New Zealand Government did not wholly favour this separation as it had always believed that New Zealand and Australia

Parachute handling was part of aircrew training, while parachute packing was a specialist ground trade in the RNZAF.

Vickers Vincents at Omaka in 1942.

should be viewed as one strategic entity, but it recognised the urgency of finalising an arrangement and so acquiesced.

On the day that Japan entered the war the RNZAF's aircraft on the strength of first-line units comprised 24 Lockheed Hudson bombers (a further 12 were in various stages of assembly), 42 Vickers Vincents, 20 Vickers Vildebeests, three de Havilland airliners and two Short Singapore flying-boats. Five months earlier, Britain had authorised the release of six Hudsons a month to New Zealand from RAF allocations in America.

The Hudson was an American twin-engined reconnaissance bomber, with a maximum speed of 255 miles an hour and a cruising range of 2160 miles. The Singapores, six of the Vincents and the three DH airliners were in Fiji. RNZAF Detachment Fiji, often referred to as 'Unit 20', had been established there in November 1940 under Squadron Leader D. W. Baird, to assist the Army in the defence of Fiji and to give what protection it could to shipping. The unit was equipped initially with four DH89 Dragon Rapides and one DH60 Moth for communications flying. When two Rapides were destroyed in a hurricane, two DH86 replacements were provided. (One of these aircraft was being overhauled in New Zealand when the Pacific War broke out.)

Aerodromes were under construction at Nandi*, on the west coast of Viti Levu, and at Nausori, 15 miles from Suva, the capital, while at Suva an RNZAF Headquarters had been established. The aircraft were operated by the Detached Flight Nandi on the west coast, under Squadron Leader G. R. White. This flight was upgraded in status on 8 October 1941 and named No. 4 (General Reconnaissance) Squadron.

Two further Singapore flying-boats were ferried from Seletar, Singapore, during December 1941 — a long flight with obsolescent aircraft, fraught with some risk from mechanical failures. The flying-boats, which first operated from Suva Harbour while a breakwater was being constructed at Lauthala Bay, were formed into No. 5 (GR)

* Note that the RNZAF uses the forms Nandi and Lauthala Bay rather than the Fijian spelling, Nadi and Laucala.

A fine closeup photograph of a Lockheed Hudson Mk III bomber many of which served with No. 2 (GR) Squadron at Nelson.

A Hawker Hind of No. 20 (Army Co-operation) Squadron, at RNZAF Station Onerahi (Whangarei), in early 1943.

Below: Avro Ansons went into service as navigation trainers in 1942, first at Omaka with the School of General Reconnaissance and then at New Plymouth when the school was transferred there later that year. Twenty-three Ansons were purchased. They were a familiar sight over Taranaki for the last three years of the war.

Blackburn Baffins of the Auckland Territorial Squadron over Westhaven, Auckland. *Fred Taylor*

A flight of early North American Harvard trainers over Wigram in February 1942.

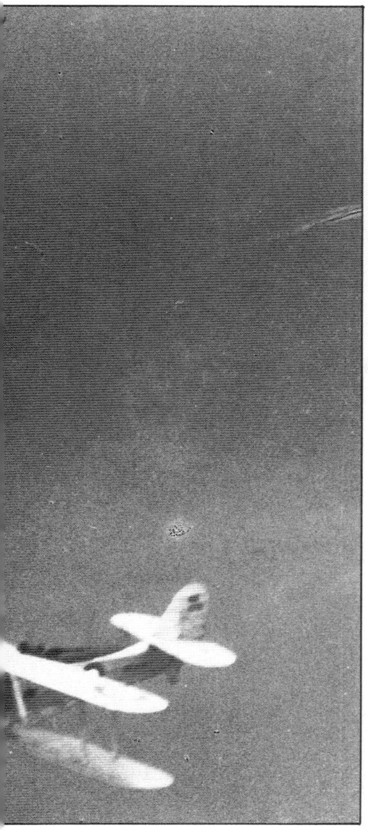

A Short Singapore flying-boat of No. 5 Squadron being slipped from the ramp in Fiji.

A Short Singapore flying-boat over Fiji.

Squadron. This unit began operations in January 1942 and, equipped with three Singapores (the fourth had been damaged on a reef in December), was employed on shipping escorts and long-range anti-submarine patrols.

Second-line aircraft within the RNZAF at the start of 1942 — mostly used for training but available for operational use in an emergency — comprised 62 North American Harvards, 142 Airspeed Oxfords, 30 Fairey Gordons, 221 de Havilland Tiger Moths, 46 Hawker Hinds, six Vildebeests, six Vincents, one Walrus amphibian and 16 miscellaneous types. Few of these aircraft could be regarded as operational. The personnel strength of the RNZAF at home was under 11,000. In desperation, New Zealand appealed to Britain for help. But the 'Europe first' grand strategy dictated the reply; apart from the few Hudsons already sent, Britain could promise no further aid.

Shortly before Japan entered the war, a conference held in Suva between New Zealand and American officers and the Fijian Government had agreed that the aerodrome at Nandi should be further developed and extended to serve the Americans as a base and another strip be built nearby at Narewa. These tasks were undertaken by No. 2 (Aerodrome Construction) Squadron, RNZAF, together with a 1000-strong Civil Construction Unit assembled by the then Public Works Department in New Zealand, and the Fijian Public Works Department. In May 1942, after the Civil Construction Unit had withdrawn to New Zealand, the RNZAF Aerodrome Construction Squadron completed the aerodrome and on 18 July the command of all armed forces in Fiji, and responsibility for its defence, passed to the United States Army.

As the situation in the Pacific deteriorated, New Zealand repeated her demands for aircraft. Medium bombers were an urgent requirement; Hurricane fighters were wanted for operational training and Army support. Again Britain refused. New Zealand then asked for complete squadrons, long-range fighters and interception fighters. The British Government replied that it could not send the squadrons but it would release to Australia and New Zealand 142 Curtiss P-40 Kittyhawk fighters from RAF orders placed in the United States and destined for the Middle East. After consultation with Australia, 18 Kittyhawks were promised to New Zealand, for delivery in March-April 1942. Later the RNZAF was allotted 36 Hudsons and 80 more Kittyhawks for delivery between March and May.

To meet the Japanese threat the flying training schools had to be reorganised. In February 1942 the Initial Training Wing, which had earlier been established at Levin, was moved to Rotorua, leaving Levin free to accommodate the Bomber Operational Training Unit. No. 3 SFTS at Ohakea was disbanded and the other SFTSs at Wigram and Woodbourne were enlarged. No. 4 EFTS at Whenuapai was disbanded in March so that No. 2 EFTS at New Plymouth was now the only flying training school remaining in the North Island. This trend to concentrate operational flying in the North Island and training in the South Island set a pattern that was largely to be maintained.

From 1942 onwards an increasing number of recruits for the RNZAF came from the Air Training Corps. The

Army co-operation, 1942.

Heavily defended Rabaul is the target for this Ventura of No. 2 (Bomber Reconnaissance) Squadron early in 1944.

A Catalina of No. 6 (Flying-Boat) Squadron at Halavo Bay, Florida Island.

Corsair strike aircraft taking off from Piva airstrip, Bougainville, with active volcano Mount Bagana in the background.

Corsair fighters being serviced at RNZAF Field Headquarters, Emirau, 1944.

Ground staff with bomb train, No. 2 Squadron, Nelson.

establishment of an Air Cadet organisation had first been suggested by the Governor-General of New Zealand, Lord Galway, in August 1940 and, following a manpower review in relation to aircrew requirements, the War Cabinet had authorised the formation of an Air Cadet Training Corps in February 1941.

The name of the organisation was finally shortened to Air Training Corps and membership was offered to boys aged 16½ to 18 years. The first six squadrons were formed in Wellington, Auckland and Dunedin during September and October. The ATC then contained two types of unit — town and school. In some town and country areas, because of the small numbers offering, units were not actually formed but cadets were enrolled and trained by correspondence courses. In March 1942 the age limits were widened. In town units boys aged 16 to 18 could enrol and in school units boys could be accepted provided they reached the age of 15 during the school year. By June 1942 there were 22 town units and 42 school units, comprising an overall total of 5000 cadets. (At peak strength in mid 1944 cadet numbers stood at 9294. By the war's end the ATC had provided 3500 enlistments in the RNZAF.)

When the defences of Fiji were being strengthened in 1940–41, several new squadrons had been formed in New Zealand to meet the growing threat of war in the Pacific. The original New Zealand General Reconnaissance Squadron at Whenuapai (which had been formed by an amalgamation of three of the territorial GR units) became known as No. 1 Squadron. It was commanded by Squadron Leader G. N. Roberts. No. 2 Squadron, equipped with Vincents and Vildebeests, was formed at Nelson in January 1941 under the command of Squadron Leader R. J. Cohen and No. 3 Squadron, flying Baffins* and commanded by Squadron Leader C. L. Monckton, was formed at Harewood in April 1941.

The squadron at Nelson, besides carrying out its own training and operational programmes, also developed a School of General Reconnaissance and pilots from the other squadrons were attached to it for training. By the end of 1941 the school had become a separate unit and it was transferred to Omaka near Blenheim. By late 1941 Nos. 1, 2 and 3 Squadrons were partially equipped with the new Hudson bombers, and soon after Japan entered the war No. 4 Squadron in Fiji was issued with six Hudsons.

Early in 1942 the RNZAF began to create new operational squadrons. In February No. 6 (Army Co-operation) Squadron was formed at Milson, Palmerston North, equipped with Hinds. In the same month No. 7 (GR) Squadron was formed at Waipapakauri, with Vincent and Vildebeest aircraft, under Squadron Leader A. J. Turner. No. 8 (GR) Squadron was formed at Whenuapai in March, with Vincent and Vildebeest aircraft, under Squadron Leader C. L. Monckton, and moved almost at once to Gisborne. No. 14 (Fighter) Squadron was formed at Ohakea in April, with Harvard aircraft, moving the following month to Masterton under the command of Squadron Leader J. N. MacKenzie, with survivors of No. 488 Squadron, pilots and ground staff, as its nucleus and equipping there with Kittyhawk aircraft.

In June No. 15 (Fighter) Squadron formed at Whenuapai, with Harvards and Kittyhawks, commanded by Squadron Leader A. Crichton, who was to take the unit to Tonga in October. No. 16 (Fighter) Squadron was formed at Fairhall, near Blenheim, in June, commanded by Squadron Leader A. N. Jones and equipped with Harvards and Kittyhawks. In July a detached flight from No. 4 (GR) Squadron stationed at Plaine des Gaiacs, in New Caledonia, was redesignated No. 9 (GR) Squadron and commanded by Squadron Leader D. E. Grigg.

New Army Co-operations Squadrons were formed. No. 20 (AC) Squadron, under Squadron Leader (acting Wing Commander) J. G. Fraser, RAF, was formed at Onerahi, near Whangarei, in August, equipped with Hinds; No. 6 (AC) Squadron was disbanded to become No. 21 (AC) Squadron in October under Squadron Leader S. N. Wiltshire, GC; and No. 22 (AC) Squadron was established at Norwood (Canterbury) with Vincent aircraft, under Squadron Leader W. A. Cameron in November. The AC Squadrons were all short-lived and none was destined to get modern aircraft or see service overseas.

Hudson turret.

* The surviving Baffins were scrapped later in 1941.

In October another fighter squadron, No. 17, was formed at Ohakea with Kittyhawks. Commanded by Squadron Leader J. V. A. Reid, it was transferred to Whenuapai in November and to Seagrove in December.

Alarmed at the possibility of Japanese troop transports appearing off the New Zealand coast, the RNZAF early in 1942 introduced a scheme whereby training units could, in an emergency, be used operationally. This was known as FAFAI (Forces Available for Anti-Invasion) and the aim of the project was simply that in an invasion every aircraft could be used offensively. To this end, auxiliary squadrons were formed: a bomber-reconnaissance squadron of 12 Vincents at the School of General Reconnaissance, another of 18 Oxfords and a fighter-bomber squadron of 12 Harvards at each of the service flying training schools, a light bomber squadron of 18 Tiger Moths at each of the elementary flying training schools, a transport squadron consisting of the civil aircraft of Union Airways and a composite squadron of Oxfords, Harvards and Tiger Moths at the Central Flying School which had been formed at Tauranga.

All aircraft in the training schools were modified to enable them to attack shipping and to strafe enemy troops. Even Tiger Moths were fitted with racks for light anti-personnel bombs. Later in the year, no invasion having taken place, the FAFAI scheme was abandoned.

Up to the outbreak of war the RNZAF Stores Depot had been located at Hobsonville. RNZAF Station Wigram acted as a depot, too, but to a lesser degree. Expansion of the Air Force on the flying side called for a parallel expansion of its stores facilities so in 1940 No. 2 Stores Depot was established at Rongotai (Wellington) when the temporary buildings erected for the Centennial Exhibition became vacant. When Japan entered the war No. 1 Stores Depot moved from Hobsonville to Hamilton, while No. 2 Stores Depot was transferred from Rongotai to the Mangaroa Valley, north of Upper Hutt, where an RNZAF Station was established. No. 3 Stores Depot was formed at Weedons, 11 miles from Christchurch, after being briefly located in buildings in that city, and No. 4 Stores Depot was established at Te Awamutu.

Wing Commander J. D. Hewett, AFC, the first officer to command No. 42 Squadron. Before this unit achieved squadron status, he commanded it as the RNZAF Communications Flight, Rongotai.

With the arrival in 1942 of new types of aircraft there was clearly a need to expand the RNZAF's repair and maintenance organisation. Prewar planning had sited the sole Repair Depot at Hobsonville. The expansion that accompanied implementation of the British Commonwealth Air Training Plan in New Zealand required all FTSs to carry out their own overhauls and by the end of 1940 all had their own engine and airframe repair shops. The EFTSs also carried out their own complete overhauls until the de Havilland factory at Rongotai was able to undertake the work.

Commanding Officers' Conference, RNZAF Ohakea, April, 1942.

Back row, left to right: Squadron Leader C. L. Monckton, Squadron Leader I. E. Rawnsley, MBE, Squadron Leader R. A. Kirkup, Wing Commander M. F. Calder, ADC, Squadron Leader A. C. Upham, DFC, Squadron Leader A. C. Parsons, Wing Commander B. S. Nicholl, Wing Commander G. L. Stedman, Wing Commander K. L. Caldwell, MC, DFC, Wing Commander R. J. Cohen, AFC, Wing Commander P. E. G. G. Connolly, RAF.

Middle row, left to right: Squadron Leader A. J. Turner, Squadron Leader E. Smart, Wing Commander R. W. K. Stevens, RAF, Wing Commander F. R. Dix, Group Captain G. N. Roberts, AFC, Wing Commander R. J. Sinclair, Wing Commander Sir Robert H. Clark-Hall, KBE, CMG, DSO, Wing Commander J. Seabrook, AFC, ADC, Group Captain M. W. Buckley, MBE, Group Captain G. S. Hodson, AFC, RAF, Wing Commander J. G. Fraser, RAF.

Front row, left to right: Mr Turner, Wing Commander E. A. Gibson, Group Captain S. Wallingford, ADC, Group Captain A. de T. Nevill, CBE, Hon. F. Jones, Minister for Defence; Air Commodore R. V. Goddard, CBE, RAF, Chief of Air Staff; Group Captain F. E. T. Hewlett, DSO, OBE, Hon. A. Hamilton, MLC, Mr T. A. Barrow, Air Secretary; Group Captain J. L. Findlay, MC, Wing Commander H. G. L. Allsop, RAF.

RNZAF Station Whenuapai in March 1941. Concrete runways were added in 1942.

RNZAF Station Ohakea in January 1941.

RNZAF Station Woodbourne in July 1941. In addition to numerous Vincents and Oxfords, this view also includes some of the recently arrived Harvards.

Camouflaged hangars and dormitories at Wigram in February 1942.

A prewar view of RNZAF Station Hobsonville.

In early 1942 steps were taken towards centralisation. While Hobsonville became purely an assembly depot, No. 1 Repair Depot was established at Rukuhia, near Hamilton, where an airfield had been constructed, and began operating there in September. No. 2 RD was established at Ohakea in June and No. 3 RD was formed in Christchurch (the airframe repair shop located at Harewood and the engine repair shop in the city), commencing operations early in 1943.

The year 1942 also saw facilities upgraded. Concrete runways at Ohakea and Whenuapai were authorised by Cabinet in January and work was completed by June. The airfield at Hobsonville was extended; airfields were developed for the new squadrons and camps created. In March £245,000 was allocated for camouflaging airfields. The Government also decided to construct dummy aircraft. By mid 1943, when work ceased on this project, 254 Hudsons, Kittyhawks and Hurricanes had been produced.

In the crucial year of 1942 the RNZAF had to face up to three major commitments. First, it had to maintain the promised output of aircrew trainees required under the Empire Air Training Scheme; second, it had to develop an organisation capable of playing its part in the defence of New Zealand; and third, it had to prepare for a contribution of operational squadrons for the war in the Pacific. The demands of these three programmes called for a rapid increase in numerical strength. Between the beginning of January and the end of December 1942, more than 16,000 new recruits were enlisted.

First day on station for a new intake of airwomen.

Wing Officer Frances I. Kain (left), the first superintendent of the Women's Auxiliary Air Force, briefs members of the first contingent of WAAFs to serve in Fiji before their departure from New Zealand.

Women played an important part. The Women's Auxiliary Air Force (WAAF) was formed in 1941 with a view to releasing more men for overseas service. In the initial stages it met with a degree of masculine prejudice but if there was resentment at the intrusion of women in some quarters there was plenty of support where it mattered most — at the command levels. At RNZAF Station Harewood, in particular, the first WAAFs had a firm champion in the Commanding Officer, Wing Commander Sir Robert Clark-Hall.

The first draft, of 200 WAAFs, was posted to RNZAF Station Rongotai in April 1941. At first airwomen were not given Service ranks, but in 1942 the WAAF was incorporated as part of the RNZAF, and women held ranks equivalent to the men's and wore similar rank badges. By late 1942 airwomen were serving on 21 RNZAF stations. (Subsequently, they also served in Fiji and on Norfolk Island.) Commissions were offered to women serving as supervisors or on cipher duties.

The first Superintendent of the WAAF was Mrs Frances I. Kain. Her assistant was Mrs E. N. Carlyon, who duly succeeded her in office. At first women were employed as cooks, stewards (then called mess hands), drivers, clerks, equipment assistants, medical orderlies and shorthand typists, but as the war progressed other trades were opened to them. At its peak strength, in July 1943, the WAAF numbered more than 3600 women, about 100 of whom were commissioned officers.

Parachute packing.

CHAPTER TEN

Action in the Pacific

The Japanese Air Force suffered an almost total and relatively rapid eclipse in the Solomons largely because of the vast resources of the United States and the strategy behind their deployment. In assessing the RNZAF's contribution to this eclipse, it must be remembered that its squadrons operated under American control and therefore had specific tasks allotted to them, tasks they discharged in a manner which earned the highest respect of their American allies.

From the time they entered the Pacific War until the opening of the Second Front in Europe, RNZAF squadrons destroyed 104 Japanese aircraft, of which the RNZAF Fighter Wing accounted for 99. This figure may not appear large, but it must be remembered that most of the major air engagements were fought by large concentrations of American aircraft with relatively small numbers of New Zealanders operating in support.

Thirteen RNZAF fighter squadrons fought in the Pacific War, some of them completing several tours of operations. Six bomber squadrons, two flying-boat squadrons, one dive-bomber squadron and two torpedo-bomber* squadrons also saw action in the Pacific campaign. How did the RNZAF come to contribute as many as 24 squadrons to this war theatre — a transport squadron staging from New Zealand to Pacific bases actually brought the total to 25 — and where did the aircraft come from?

When New Zealand was made a part of the American South Pacific Command in 1942 and Admiral Robert L. Ghormley, USN, arrived in May to set up his headquarters in Auckland, it was assumed that new equipment for the RNZAF would quickly be arranged. At his first meeting with the New Zealand Chiefs of Staff Committee a few days later, however, Ghormley confounded their expectations, bluntly stating that equipping the land and air forces of New Zealand was no part of his responsibility. As the New Zealand Army was being adequately supplied by Britain, the Army authorities were not unduly perturbed, but for the Air Force the statement came as a serious blow.

Negotiations between Britain, New Zealand and the United States followed, New Zealand contending that her air force must be urgently equipped not only for the defence of her home territories but also to take part in offensive operations against the Japanese in the Pacific. New Zealand proposed that her air force be expanded to a strength of 20 squadrons by April 1943 and that certain of these units should be used operationally against the enemy. This proposal was submitted to the Joint Planning Staffs in both the United Kingdom and the United States. While it was being debated an agreement known as the Arnold-Slessor-Towers Agreement (or the Arnold-Towers-Portal Agreement)* was signed by Britain and the United States in Washington. This document required America to define and satisfy the strategic requirements of Australia and New Zealand.

The American Joint Planning Staff recommended that the RNZAF be limited until April 1943 to a strength of 10 squadrons — five fighter squadrons equipped with Kittyhawks, four light bomber squadrons flying Hudsons and one Army co-operation squadron. The British Chiefs of Staff, however, recommended the formation of a further six squadrons by the RNZAF taking over and manning six American squadrons in the Pacific — fighters and B25 bombers.

In August Air Commodore R. V. Goddard, RAF†, who had succeeded Air Commodore Saunders as Chief of Air Staff the previous year, flew to Hawaii and then to Washington to use his influence in negotiating a positive combat role for the RNZAF. He saw Admiral Nimitz at Pearl Harbour, revealed the strength of the RNZAF in operational terms and asked for modern aircraft.

Admiral Nimitz referred him to Washington where he proceeded the following day. There he told the Chiefs of

Air Commodore R. V. Goddard, CBE, RAF, the RNZAF's third Chief of the Air Staff.

* They did not drop torpedoes, however, but were used mainly as light bombers.

* General H. H. Arnold, Admiral J. H. Towers and Air Marshal Sir John Slessor, the last named being succeeded by Air Marshal Sir Charles Portal.

† Air Marshal Sir Victor Goddard, KCB, CBE, DSM(US), RAF.

Flight Lieutenant W. R. B. Watson, B Flight commander of No. 3 Squadron at Guadalcanal, was the first member of the RNZAF to receive the American DFC in the Pacific War.

Staff of his plan to expand the RNZAF to 25 squadrons in 18 months with the help of the United States. He returned home in the certain knowledge that the 10-squadron plan would proceed forthwith and with a personal assurance from Admiral E. King, the American Naval Commander-in-Chief, that 25 squadrons would be equipped but that this would take two years to achieve.

Air Commodore Goddard had played his cards well. In fact, the RNZAF had 24 operational squadrons in service within two years, while one of its two transport squadrons based in New Zealand flew regularly to RNZAF Pacific Island bases carrying men and supplies.

The Mutual Aid Agreement, always commonly referred to as the Lend-Lease (or Lease-Lend) Agreement, was signed in Washington by the US Secretary of State, Mr Cordell Hull, and Mr Walter Nash, New Zealand's Minister of Finance, on 3 September 1942. The contracting nations pledged their full resources, military and economic, against those countries with whom they were at war; it was similar to an agreement concluded between the United States and Great Britain earlier that year. Under Lend-Lease, as much as possible of each government's contribution was to be in the form of reciprocal aid.

The Allied cause in the Pacific was now rallying. American Marines landed on Guadalcanal in August 1942 and fierce air, sea and land battles followed. But the enemy still had control of the sea, and to a great extent, the air, in the Southern Solomons.

The RNZAF, with a clear-cut role in the Pacific campaign defined before the battle for Guadalcanal, sent a force of Hudsons to New Caledonia under Squadron Leader C. J. Kidson, at the request of Rear Admiral McCain, Commander Air, South Pacific, who was short of aircraft for anti-submarine patrols. From this force of four officers and 30 other ranks the unit grew quickly to a strength of 36 officers and 245 other ranks. It was then designated No. 9 Squadron, RNZAF, and placed under the command of Squadron Leader D. E. Grigg.

The battle for Guadalcanal was essentially a battle for the airfield which the Americans needed to hold — they had captured it in August — if they were to achieve adequate air support for their naval and land forces. For a fortnight after the original landing on 7 August, the US Marines had little opposition on land, but the disastrous naval action off Savo Island on the night of 9 August, in which a Japanese force sank four American and Australian cruisers, had given the enemy full, if temporary, control of the sea. Japanese naval units were thus able to bombard the Marines' positions ashore at will and as the enemy sought to overwhelm the Americans there was much heavy land fighting. The Japanese made their last major attempt to land reinforcements with naval support in the second week of November.

The RNZAF joined forces with the Americans in this theatre on 23 November. The first New Zealand unit to operate from Guadalcanal was No. 3 (Bomber Reconnaissance) Squadron which was sent from New Zealand as a complete unit numbering 300 men under the command of Squadron Leader G. H. Fisher. The squadron's 13 aircraft, led by Fisher, went first to Espiritu Santo via Norfolk Island and New Caledonia and the main party comprising 11 officers and 218 other ranks was transported by the Union Steam Ship Company's passenger steamer *Wahine*. Another vessel, the *Taybank*, followed with a small party of airmen and 387 tons of equipment. Leading this party was Pilot Officer W. A. Chandler who, since his return from the Far East, had been commissioned as an engineer officer.

Between October and December No. 3 Squadron operated from Pallikulo airstrip, Espiritu Santo, flying daily air searches and anti-submarine patrols around the fleet anchorage in the Segond Channel. For the last fortnight of this period six of the squadron's Hudsons and eight crews were detached to Guadalcanal, part of which was still occupied by the Japanese. The Hudsons had been preceded by a small servicing party, which had travelled by sea and set up camp in a jungle-clad valley close to

Squadron Leader G. H. Fisher (second from left, holding pencil), Officer Commanding No. 3 Squadron, briefs a Hudson bomber crew at Guadalcanal for a mission. Studying the map (second from right) is Flying Officer G. E. Gudsell, pilot, the first member of the RNZAF to receive a decoration in the Pacific War. At far left is Flight Lieutenant A. C. Allen, A Flight commander, at centre Flying Officer J. F. Gabites, meteorological officer, and at the far right Flight Sergeant R. Doak, wireless operator in Gudsell's crew. One crew member is obscured. Fisher's command of No. 3 Squadron was recognised by the award of the US Air Medal. *New Zealand Herald*

Henderson Field, which was codenamed 'Cactus'. The primary role of 3 Squadron was to search and patrol and it was attached to the Search and Patrol Group of the Air Search and Attack Command, which was part of US Task Force 63.

The Hudson was essentially a reconnaissance aircraft; it was armed with machine guns for self-defence and carried bombs, but it was impressed on the crews that their job was to report enemy movements and not to go looking for action.

Sometimes, however, they had action thrust upon them and they acquitted themselves well. The advance flight of 3 Squadron was on operations the day after it arrived on Guadalcanal. A Hudson captained by Flying Officer G. E. Gudsell spotted a tanker and two transports escorted by a destroyer to the south of Vella Lavella. Gudsell's plane was engaged by three Nakajima floatplanes which were circling the convoy but they did not press home their attacks, probably deterred by the Hudson's armament. The engagement lasted 20 minutes and no direct hits were scored.

A more serious engagement occurred on 27 November when Gudsell sighted an enemy task force to the south-west of Vella Lavella. Gudsell reported the composition of the force, and then closed with it to make a better inspection. As he was doing this he was dived on by three land-based Japanese fighters. In their first attack they put the Hudson's top turret gun out of action, and then concentrated on attacks from astern. Gudsell directed the action from the astro-hatch while the second pilot, Flying Officer R. M. McKechnie, flew the aircraft. After an action which lasted 17 minutes the Japanese aircraft broke off and retired, having scored only three hits and without having injured any of the Hudson's crew. This early action contributed to the high morale which prevailed in the squadron throughout its tour of operations.

In the previous month the Americans had lost a number of their search planes through enemy action and after seeing the comparatively light armament of the Hudsons, had told the New Zealanders they would be 'sitting ducks' for Japanese fighters. The proof that, when properly handled and resolutely fought, the Hudson could repel odds of three to one was comforting to all the aircrews. These incidents resulted in the award of the US Air Medal to Flying Officer Gudsell, who thus became the first member of the RNZAF to win a combat decoration in the Pacific War.

Throughout December A and B Flights served in turn as the advanced element of the squadron at Guadalcanal. The crews flew for long stretches over water, often without seeing anything of significance, but they had to be constantly alert, not only for likely targets but also for enemy fighters. Japanese shipping was usually too strongly defended and they could only radio back to base and call up American strike forces to deal with the enemy. This they frequently did most effectively and earned high praise from their American superiors. Occasionally, they found a target which they could attack unaided.

On 2 December a Hudson captained by Sergeant I. M. Page was on a routine patrol to the west of New Georgia when it sighted, some 12 miles away, what at first looked like a canoe with a native standing in it. Approaching to investigate, Sergeant Page identified the object as a submarine of 500 tons, fully surfaced, with a lookout on the conning tower. The Hudson made a bombing run out of the sun and dropped two 250 lb anti-submarine bombs and two 325 lb depth charges as the submarine crash dived. Although three of the bombs fell short, one landed over the target and a large patch of oil came to the surface. The submarine was almost certainly damaged, but in the absence of positive proof no claim was made of its destruction. For this action, Sergeant Page was awarded the DFM.

Guadalcanal was a depressingly primitive spot. There was mud everywhere and the heat was truly formidable, more oppressive than on most other islands where the RNZAF fought. The camp consisted solely of tents erected over the mud and the absence of dugout accommodation meant that during air raids the men had to tumble out of bed into the nearest foxholes. It was not surprising that broken nights and loss of sleep caused fatigue, but if the efficiency of 3 Squadron was threatened by it, it was not impaired. Mosquitoes abounded; in the valley where the New Zealanders lived the air was thick with them and many crew members developed malaria.

Hudson bombers of No. 3 Squadron lined up on the tarmac at Whenuapai before departing for Pacific service.

On 14 January 1943 Squadron Leader H. C. Walker, AFC, arrived at Guadalcanal to assume on the spot command of 3 Squadron. By this time the complete squadron was at Guadalcanal and the camp in which it had lived at Espiritu Santo had been reorganised as an RNZAF station with Fisher, newly promoted to acting wing commander, in command. Wing Commander Fisher retained his overall command of 3 Squadron, while Walker led the unit in the field.

Squadron Leader Walker had been responsible for converting pilots on to Hudson aircraft in New Zealand in 1941 and on 9 December — two days after Pearl Harbour — he had delivered the first Hudsons to Fiji and remained there with No. 4 Squadron throughout 1942, the year he had been awarded the Air Force Cross. He was, therefore, particularly qualified to command a bomber squadron operationally. Walker lost no time in establishing a close liaison with Brigadier General F. P. Mulcahy, commanding the 2nd Marine Air Wing on Guadalcanal, through Colonel Brice, and had soon earned the respect of the American command.

On 4 February 1943 3 Squadron carried out its first night bombing. Flight Lieutenant J. S. Knobloch stood down to allow Squadron Leader Walker to captain Hudson NZ2033 with Pilot Officer John Coom as second pilot and Flight Lieutenant H. N. Blundell as navigator. Group Captain Wallingford also flew on this mission. The target was a radar station on West Cape, Choiseul Island, and the mission was successful. On the return flight the crew observed five Zeros 1000 feet below them, flying into the dawn, but the enemy did not see the Hudson.

Squadron Leader Walker led No. 3 Squadron on Guadalcanal from 20 January until 3 April 1943, then returned to New Zealand to take command of No. 1 (BR) Squadron. The manner in which he led 3 Squadron in those crucial months is reflected in the award by the American authorities of the US Legion of Merit (Degree of Officer) which was conferred on Walker on 9 August 1943.

The first engagement of No. 3 Squadron, south of Vella Lavella.

Squadron Leader H. C. Walker, officer commanding No. 1 Squadron, confers at Guadalcanal with Air Commodore S. Wallingford (left) and Air Commodore M. W. Buckley (right) in November 1943. Air Commodore Wallingford was then AOC No. 1 (Islands) Group and Air Commodore Buckley was about to succeed him in this post.

Fighter Squadrons in Action

Although No. 14 Squadron was the first fighter squadron to be formed for Pacific combat duties, No. 15 was the first to go into action. After serving in Tonga for six months under Squadron Leader A. Crichton, it moved to Espiritu Santo in February 1943 and to Guadalcanal in April. (In Tonga the squadron had taken over the aircraft of No. 68 Pursuit Squadron USAAF. The later fighter squadrons went to the forward area equipped with their own aircraft.)

Crichton had two seasoned pilots as his flight commanders. Flight Lieutenant M. J. Herrick, although only 20, was already a veteran of the Battle of Britain and Flight Lieutenant J. A. A. Gibson had fought in both the Battle of Britain and the Battle of France. Herrick had destroyed seven enemy aircraft and Gibson was credited with 13½. Gibson had baled out on three occasions and both men had been decorated with the DFC.

When Crichton was killed in a flying accident at Nandi, Fiji, during a final working up period there, command passed to Herrick who established the squadron at its new home on Guadalcanal. Herrick was to be the first New Zealander to destroy an enemy aircraft in the South-West Pacific campaign, sharing this excitement with Flight Lieutenant S. R. Duncan on 6 May 1943.

Aircrew members of No. 3 Squadron at Guadalcanal, December 1942. Back row, from left: Flying Officer Steve Stevenson, Flying Officer J. F. Gabites, Flying Officer J. J. McDowell, Flying Officer W. J. S. Monks, Flight Lieutenant A. C. Allen, Flight Lieutenant W. R. B. Watson, Flying Officer J. Coom, Flying Officer L. Hood. Middle row: Flying Officer R. M. McKechnie, Flying Officer A. N. Arnott, Flying Officer H. C. Carter, Flying Officer B. E. Malone, Flight Lieutenant E. A. Barker, Flying Officer J. S. Knobloch, Flying Officer R. J. Alford. Front row: Flying Officer R. W. Walker, Flying Officer G. E. Gudsell, Flying Officer W. D. H. Smith, Flying Officer A. B. Horrocks, Flying Officer S. D. Walker.

The two pilots were escorting a Hudson on patrol in an area where enemy planes had been recently sighted. The Hudson was flying at 500 feet with the two fighters at 3000 feet and slightly to the rear. An enemy floatplane was sighted from the astro-hatch of the Hudson about 3 miles away, at a height of about 1500 feet. Neither of the fighters had seen it so the Hudson set off in pursuit but lost the enemy in a heavy rainstorm. The fighters then picked him up. The Hudson's crew heard their 'tally-hos' over the R/T and saw them jettison their belly tanks, give chase and then open fire. The floatplane burst into flames and crashed into the sea.

The honour of shooting down the first Zero for New Zealand went to Sergeant R. A. Martin on 7 June. Martin was flying a Kittyhawk of 15 Squadron in a major air

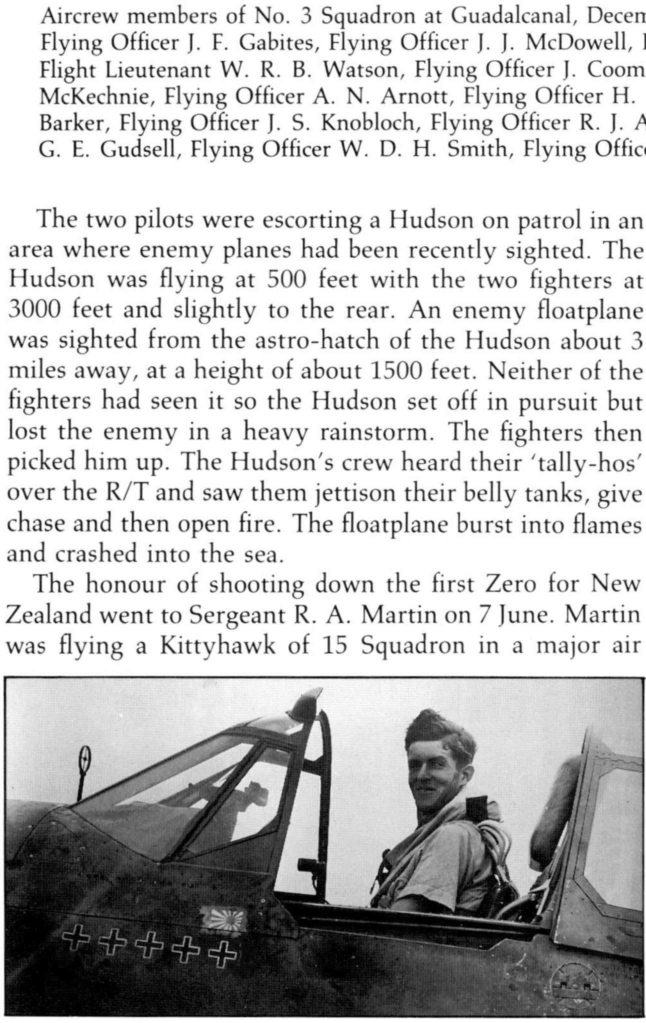

Squadron Leader M. J. Herrick, DFC.

battle in which Allied fighters claimed 23 out of a force of 40 to 50 enemy aircraft. At 23,000 feet the 12 Kittyhawks intercepted the enemy over the sea just south of Russell Island. Zeros were coming in from all directions and the New Zealanders broke formation and went in pursuit. They were in the thick of it when three Zeros attacked a section led by Squadron Leader Herrick. Both Flight Lieutenant J. R. Court and Sergeant Martin peeled off and went after one. Court's guns jammed and Martin carried on the attack. A burst from his guns penetrated the Zero's fuel tanks; flames enveloped the machine and it fell into the sea. Within 20 minutes three other Zeros had been accounted for — by Duncan, Herrick and Flight Lieutenant D. Greig in that order. American intelligence reports said afterwards that the work of the New Zealanders in this battle, in which 44 Allied aircraft took part, and which was fought in very bad weather, was outstanding.

On 22 June the squadron withdrew from Guadalcanal.

In No. 14 Squadron's first tour of operations at Guadalcanal, it shot down 22 Japanese aircraft besides four probables, losing four aircraft and three of its pilots. In that tour it was commanded by Squadron Leader S. G. Quill. Quill had been one of the first 'new' pilots to join No. 14 Squadron when it was training at Masterton under Squadron Leader MacKenzie with the survivors of 488 Squadron. (In fact, his log book for that period bears the inscription 'With 488 Squadron'; such was the reputation that the remnants of 488 brought with them to their new unit.)

Under Quill, 14 Squadron went into action within 24 hours of its arrival on Guadalcanal, shooting down six enemy aircraft for the loss of one in an engagement on 12 June 1943. The casualty was Flying Officer K. P. C. Morpeth, who was shot down in flames.

On the morning of 1 July enemy dive-bombers and fighters attacked American positions on Rendova. They were intercepted by Allied fighters, including eight aircraft of 14 Squadron led by Flight Lieutenant E. H. Brown. In the battle which followed 22 enemy aircraft were shot down for the loss of eight Allied fighters. Of these, 14 Squadron claimed seven enemy aircraft destroyed and three probables. Five pilots each accounted for one enemy aircraft and Pilot Officer R. A. Weber destroyed two. Two New Zealand aircraft were lost. Flying Officer J. F. Burton, having survived the hazards endured by 488 Squadron in the Far East, came down in the Munda area and was not seen again. Flight Lieutenant Brown abandoned his aircraft on the way back to base, landed in the water and was picked up by a rescue boat four and a half hours later.

In another action a few days later, eight aircraft of 14 Squadron were surprised by a force of more than 40 enemy aircraft while patrolling at 14,000 feet over Rendova. In this engagement Squadron Leader Quill and Sergeant R. C. C. Nairn each shot down an enemy aircraft and Flying Officer G. B. Fisken destroyed three. Quill, wounded in the shoulder, was forced to crash land at the Russell Islands, and when Nairn's aircraft was badly damaged he also landed at the Russells. Squadron Leader Quill's tour of duty as squadron commander brought him the DFC.

Formation of No. 1 (Islands) Group

No. 1 (Islands) Group Headquarters was formed at Espiritu Santo on 10 March 1943 and Group Captain S. Wallingford, who had been serving in the area as RNZAF staff officer to the Commander, Aircraft, South Pacific, Rear Admiral McCain, since October 1942, was placed in

No. 15 Squadron Kittyhawk in action.

charge, being promoted to air commodore the following month. His command embraced all RNZAF units in the South Pacific with the obvious exception of those in New Zealand. Other subordinate headquarters within the command of Rear Admiral McCain included those of 1st Marine Air Wing, 13th United States Army Air Force and United States Naval Aircraft.

In 1943 the RNZAF in the South Pacific did not act as one strategic or tactical force with a specific mission. Individual squadrons in the forward area came under the operational control of the local American commander. At the same time, administrative control of its own units was exercised by the Air Department in Wellington through the establishment of No. 1 (Islands) Group.

To understand properly the part which the RNZAF was to play in the Pacific war it is necessary to have some idea

Squadron Leader S. G. Quill.

Flying Officer R. A. Weber.

Sergeant R. C. C. Nairn.

Air Commodore S. Wallingford.

of the higher organisation of the American forces in the area. The following chart shows the chain of command:

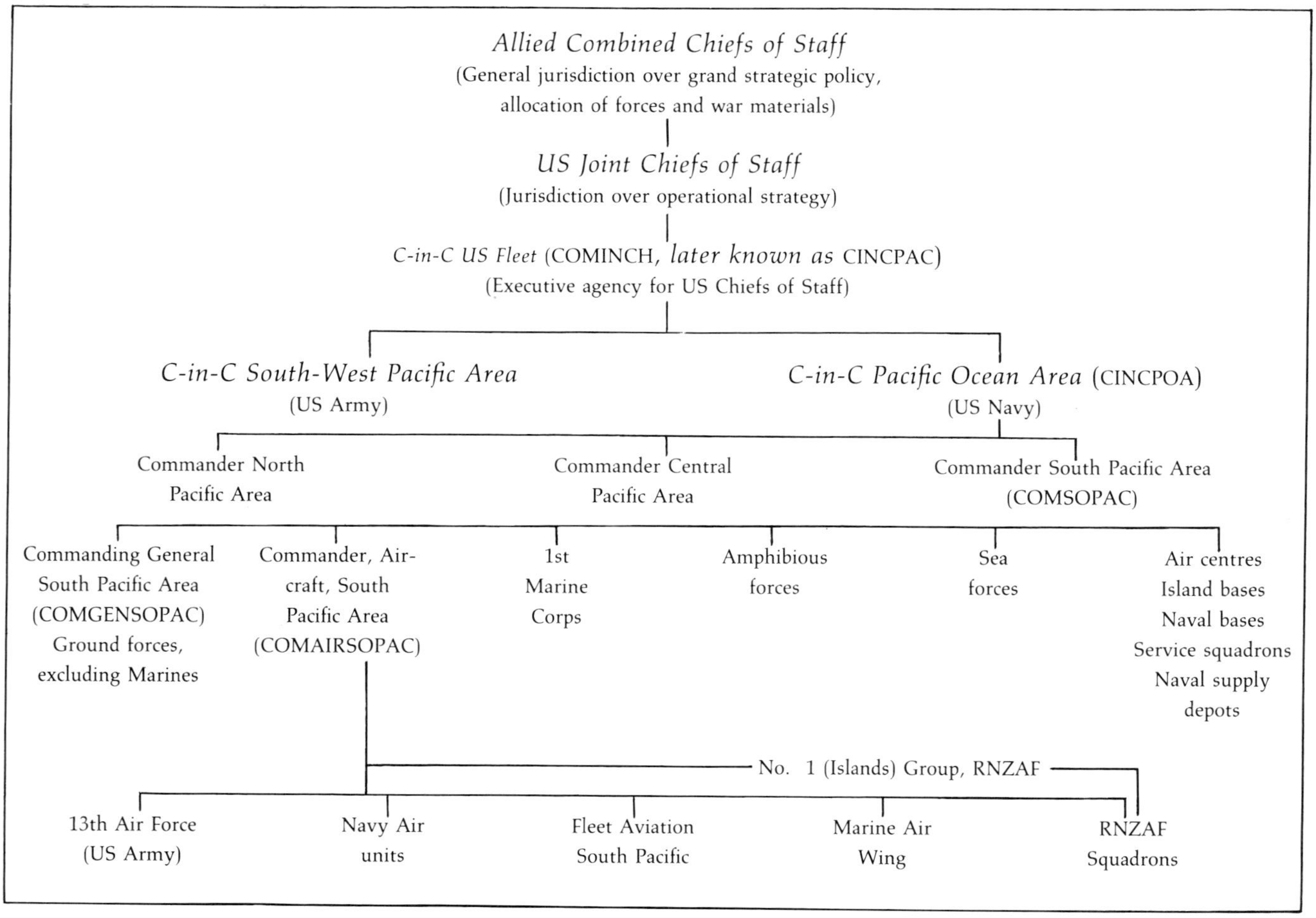

In the period immediately following the formation of No. 1 (Islands) Group the overall administration of the RNZAF units in the area was confused. No. 3 Squadron had its command headquarters at Espiritu Santo, under Fisher, while the squadron was at Guadalcanal under Walker; No. 9 Squadron's advanced elements were at Santo while its ground and headquarters organisation was still at New Caledonia, hundreds of miles distant; and Nos 14 and 15 Squadrons were in the process of moving up while their ground organisations were far behind, one in Tonga and the other in New Zealand.

The RNZAF was growing fast — some would say too fast — and inevitably much of the growth was a result of wartime expediency rather than careful planning.

The first appointments to be filled at Group Headquarters were: Senior Air Staff Officer, Wing Commander G. H. Fisher (shortly to be replaced by Wing Commander I. G. Morrison); Senior Administrative Officer, Squadron Leader J. H. J. Stevenson; Senior Medical Officer, Major J. E. Hardwick-Smith, NZMC; Senior Equipment Officer, Squadron Leader D. G. Sinclair; Group Signals Officer, Squadron Leader R. M. Kay; Group Intelligence Officer, Squadron Leader J. D. Canning (replaced in May by Flight Lieutenant T. G. Tyrer); Group Armament Officer, Flight Lieutenant W. A. Hopkins; Works Officer, Flight Lieutenant R. R. Parsons (replaced in May by Squadron Leader W. J. E. V. Grace). By the end of October the headquarters had a staff of 34 officers and 179 airmen.

Throughout 1943 the number of RNZAF units in the forward area steadily increased. By May No. 9 Squadron was at Espiritu Santo, having moved up from New Caledonia, and was providing a reserve of aircraft for No. 3 Squadron at Guadalcanal. No. 14 (Fighter) Squadron was also stationed at Espiritu Santo and No. 15 (Fighter) Squadron was on Guadalcanal. With four operational RNZAF squadrons in the area there was enough repair and inspection work to justify the setting up of an RNZAF repair depot in the forward area. This new unit was established on Espiritu Santo in June 1943 as No. 4 Repair Depot.

Until mid 1943 each RNZAF squadron was a more or less self-contained unit, having its own aircraft and maintenance staff. This system was costly in terms of ground staff and before long manpower resources were unable to sustain it. In the latter half of 1943 the policy was changed in favour of a scheme whereby land-based aircraft were allotted to a servicing unit which remained in the forward area indefinitely, its personnel being replaced from time to time after serving set periods of Pacific duty. Each squadron, consisting almost entirely of aircrew, was posted to an RNZAF station in the forward

RNZAF quarters on Guadalcanal.

On tropical island bases the visit of the YMCA vehicle was always popular.

area, where it served alongside a servicing unit (SU) for an operational tour lasting from six to eight weeks, the SU's engineer officer being responsible for providing the squadron commander daily with serviceable aircraft. At first the system was unpopular because the squadrons felt that the traditional close working relationship between aircrew and ground crew had been destroyed, but in practice the system worked quite well. This explains why so few aircraft at this time carried individual markings. At stations where two or more SUs were located, a maintenance wing was formed.

No. 52 Radar Unit had moved into Guadalcanal in the early part of 1943 and No. 1 (Islands) Works Squadron had been formed within No. 1 (Islands) Group, RNZAF, towards the end of the year. On 1 November 1943 RNZAF Base Depot, Espiritu Santo, was formed, taking over the functions of RNZAF Station, Espiritu Santo. Wing Commander I. E. Rawnsley, who had succeeded Fisher as commanding officer, handed over command to Wing Commander H. L. Tancred and took command of RNZAF Station Guadalcanal.

At its inception the Base Depot comprised No. 14 Squadron, No. 3 Squadron (which had by then returned from Guadalcanal), No. 6 (Flying-Boat) Squadron, which had been newly formed in Fiji in May and was now located at Segond Channel, No. 4 Repair Depot, Nos 1 and 12 Servicing Units and Base Depot Headquarters which included No. 2 RNZAF Hospital. The RNZAF had 1196 men in these units.

Dakota NZ3501, the first of 49 such aircraft acquired for the RNZAF in the Second World War, was flown by Squadron Leader F. J. Lucas, DFC, and was adorned by the well-known cartoon character, Popeye.

Formation of No. 40 (Transport) Squadron

Obviously the support of these formations in the South Pacific would be very largely dependent on air transport. Shipping between New Zealand and the forward area was scarce, subject to delays and beyond the direct control of the RNZAF. While much of the equipment used by the New Zealand units was shipped to island bases direct from the United States, a large quantity of materials and most personnel had to be transported to the forward area from New Zealand.

Trooping aircraft had been requested from the British Government early in 1942 and, not surprisingly, none could be spared at that time. Fortunately, the talks which Goddard had held with the Americans in 1942 were bearing fruit; six Douglas C-47 Dakotas and nine Lockheed C-60 Lodestars were allocated in 1943. Until these aircraft came to hand the RNZAF was dependent on the two United States transport organisations operating in the South Pacific — the Naval Air Transport Service (NATS) and Services Command Air Transport (SCAT), both of which had terminals in New Zealand at Whenuapai.

With American transport aircraft promised, preparations began in 1943 for the establishment of an RNZAF transport squadron at Whenuapai. A Pacific Ferry Command was formed in April under Wing Commander R. J. Cohen, who was commanding the Whenuapai station, to assist in bringing the new aircraft to New Zealand, and Squadron Leader F. J. Lucas, DFC, back home after two tours of operations in Europe, was put in charge of a one-aircraft unit (Dakota NZ3501) which would be the nucleus of the RNZAF's first transport squadron.

Squadron Leader Lucas flew this first Dakota to every island in the Pacific where RNZAF servicemen were based, and he collected valuable data for the squadron that would form up as crews delivered the new aircraft.

Dakotas for the transport squadron and Consolidated PBY Catalinas for the two flying-boat squadrons were flown from the west coast of America by New Zealand crews while Lockheed PV-1 Venturas for the bomber reconnaissance squadrons were generally shipped to Kaneohe, assembled on Ford Island in Pearl Harbour by the US Navy and then flown to New Zealand. A small RNZAF detachment was based at Ford Island to help in the assembly and to test the aircraft before they were ferried home.

Regular Dakota visits to the Islands bases were always eagerly awaited. They brought mail, and sometimes food parcels as well.

RNZAF Ventura bombers on a bombing raid in the Northern Solomons.

No. 40 (Transport) Squadron, which was to achieve a magnificent record over the years, was formed on 1 June 1943 at Whenuapai under Wing Commander J. Adams and in September that year it absorbed the detachment at Ford Island. As the new unit obtained its full quota of Dakotas and Lodestars and trained crews to fly them it was able to start operating regular schedules. One Dakota, leaving on a Monday, would fly the Whenuapai-Nausori-Santo-Guadalcanal-Santo-Whenuapai route and arrive back home on Thursday; the other, departing on Friday, flew the same circuit in the opposite direction. Two Lodestars, leaving their base on Wednesdays and Sundays, operated over the same route with an additional call at Tontouta, New Caledonia. A third Lodestar carried passengers and freight to Norfolk Island every Friday, returning the same day. A fourth, which left Whenuapai each Thursday and arrived back the following Monday, carried mail to all RNZAF bases in the South Pacific. This was the aircraft that RNZAF personnel of all ranks and trades listened for in their island huts, tents, offices, workshops and revetments — the schedule that brought letters from home. At the end of October the schedules were amended to include Ondonga (New Georgia) where the RNZAF Fighter Wing had then been established.

In January 1944 No. 1 (Islands) Group was relocated at Guadalcanal and the transfer of men and equipment to their new home was a major task for No. 40 Squadron. Air Commodore Wallingford had relinquished command in November, being succeeded as AOC by Air Commodore M. W. Buckley. Subsequently, as the RNZAF moved further forward to Bougainville, Emirau, Jacquinot Bay and Los Negros, opening stations and bringing new fighter, bomber and dive-bomber squadrons into the Pacific campaign, the tasks for No. 40 Squadron increased enormously.

The operational record of any squadron, whether in peace or war, owes much to its servicing unit. When Squadron Leader Lucas arrived at Whenuapai to begin his pioneering flights around the Pacific, Squadron Leader J. R. Torry, an engineer officer newly posted from Fiji, was forming an aircraft maintenance wing at Whenuapai, a unit that by long hours of work and high professional standards was to contribute much to the reputation of 40 Squadron.

P-40s being ferried to the forward area with a Hudson bomber escort.

CHAPTER ELEVEN

Neutralisation of Rabaul

By early 1943 the outer defences of Japan's newly acquired territories in the South and South-West Pacific were crumbling. The vast industrial power of the United States was well into its stride; new carriers had been built and sent into action and new aircraft such as the Grumman Avenger torpedo-bomber and the Chance Vought F4U Corsair fighter had entered operational service, augmenting the Kittyhawks, Dauntlesses and highly successful P-38 Lightnings which were already serving in the region. The Grumman F6F Hellcat fighter, designed in 1942 in the light of experience with the Japanese Zero, was soon to appear on the scene.

Rabaul, on the north-east tip of New Britain, had been seized by the Japanese on 23 January 1942. Its harbour and port, protected from behind by jungle-covered mountains, made it an ideal naval and air base for their campaign in the South and South-West Pacific. They had lost no time in fortifying it, building five airfields nearby and installing one of the heaviest concentrations of anti-aircraft guns in the world.

From Rabaul Japan had occupied New Guinea and the Solomon Islands but any plans to capture Port Moresby and New Caledonia and to attack the east coast of Australia had been frustrated by the Battle of the Coral Sea in May 1942 and the American landing on Guadalcanal three months later. Rabaul, however, still loomed large as a menace to the Allies and its neutralisation as a naval and air base was of paramount importance in the South Pacific campaign.

The Allies' elimination of the Japanese defence network in the Central Solomons paved the way for the final stages of the campaign designed to secure bases from which Rabaul could be attacked from the east. MacArthur's South-West Pacific forces planned to land on the western part of New Britain and so bring Rabaul into easy striking distance from the west. After reducing the defences around Rabaul, MacArthur was to push on to the Philippines.

But, because the war in Europe was considered to be of overriding importance, only some 15 per cent of the Allies' men and materials were then poised against Japan in the Pacific; General MacArthur's forces were not then sufficient to take Rabaul. MacArthur's thrusting, decisive air commander in the region, Lieutenant General George C. Kenney, turned his attention to the aerial bombardment of Rabaul as a preliminary to the invasion of Bougainville and the landing on western New Britain. As the time approached for Admiral Halsey's landing on Bougainville Kenney launched his bomber offensive against Rabaul, striking on 12 October 1943. More than 300 aircraft took part in this first raid.

The RNZAF fighter squadrons, operating first from Guadalcanal but staging through New Georgia, gave close cover to the American bombing strikes. As Allied air forces moved up in readiness for the attack on Bougainville, new airfields were added at New Georgia and Vella Lavella. On 15 October 1943 RNZAF Station New Georgia was formally established at Ondonga and Wing Commander T. O. Freeman, newly transferred from the European theatre, was appointed commanding officer. He was responsible to the AOC No. 1 (Islands) Group for co-operation with the US Fighter Command at Ondonga and for the administration of all RNZAF units in the New Georgia area.

Nos 16 and 18 Squadrons, commanded by Squadron Leaders J. H. Arkwright and J. A. Oldfield respectively, moved forward to Ondonga from Guadalcanal and became

A low-level strike by RNZAF Corsairs.

Air Vice-Marshal Sir Leonard M. Isitt, KBE, the first New Zealander to hold the office of Chief of the Air Staff.

A Sunderland flying-boat of No. 5 Squadron being serviced in the huge hangar at Lauthala Bay, Fiji.

A view of the Lauthala Bay base in the early 1960s.

known as the RNZAF Fighter Wing, with Freeman commanding both wing and station. For the first three weeks there all fighter operations were directed by a USMC colonel but on 14 November Freeman was appointed in his stead.

The triple responsibility — operations, the wing and the station — was, however, too much for one man and in December Wing Commander D. C. Campbell was brought in to take command of the station. The move to Ondonga brought the New Zealanders to within 120 miles of the Japanese bases in southern Bougainville and to less than 400 miles of Rabaul. On 27 October the first enemy air raid occurred and bombs were dropped near the New Zealand camp.

On that same day New Zealand and American troops went ashore at Stirling and Mono Islands in the Treasuries and the following morning an American parachute battalion landed on Choiseul. On 1 November Halsey landed on Bougainville, where there were about 35,000 Japanese troops. The Allies had selected Cape Torokina, in the north of Empress Augusta Bay, for the landing.

On the morning of the landing 18 Squadron was airborne and the New Zealanders witnessed the initial attack. Shortly after 8 a.m. they were vectored on to a large formation of Japanese aircraft which had been picked up by radar. Flight Lieutenant R. H. Balfour led his flight to a position 16,000 feet over Cape Torokina where they saw 50 to 60 Zeros flying down the middle of the island towards Kahili, probably unaware of the invasion below. Balfour attacked and his flight of eight aircraft shot down seven of the enemy and recorded another probable. One of the New Zealand aircraft was damaged in the engagement and had to land on the water near Vella Lavella. The pilot was picked up by a barge and rejoined his squadron two days later.

The period following the Bougainville landings was one of intense activity for the air force units on New Georgia; the RNZAF Fighter Wing achieved an outstanding record, flying more than 1000 sorties in November. In spite of heat, rain and mud, scanty, monotonous food, fatigue and air raid alarms, the morale of the New Zealanders remained high.

The development of airstrips at Empress Augusta Bay made it possible for land-based fighters equipped with long-range tanks to operate against Rabaul, both on offensive sweeps and as bomber escorts. The first attack, by 80 aircraft, was made on 17 December and Wing Commander Freeman led the RNZAF formation of 24 Kittyhawk aircraft. In an engagement with Japanese fighters Freeman's aircraft was damaged and he tried to force land on New Ireland. No trace of him was ever found.

Squadron Leader J. S. Nelson acted as CO of the Fighter Wing until Wing Commander C. W. K. Nicholls, a New Zealander in the RAF, assumed command in February 1944.

One of the wing's biggest engagements over Rabaul was the air battle of 24 December 1943. On this day the wing scored its biggest success of the war — 12 Japanese aircraft shot down, four more claimed as probables and several damaged, for the loss of five New Zealand pilots.

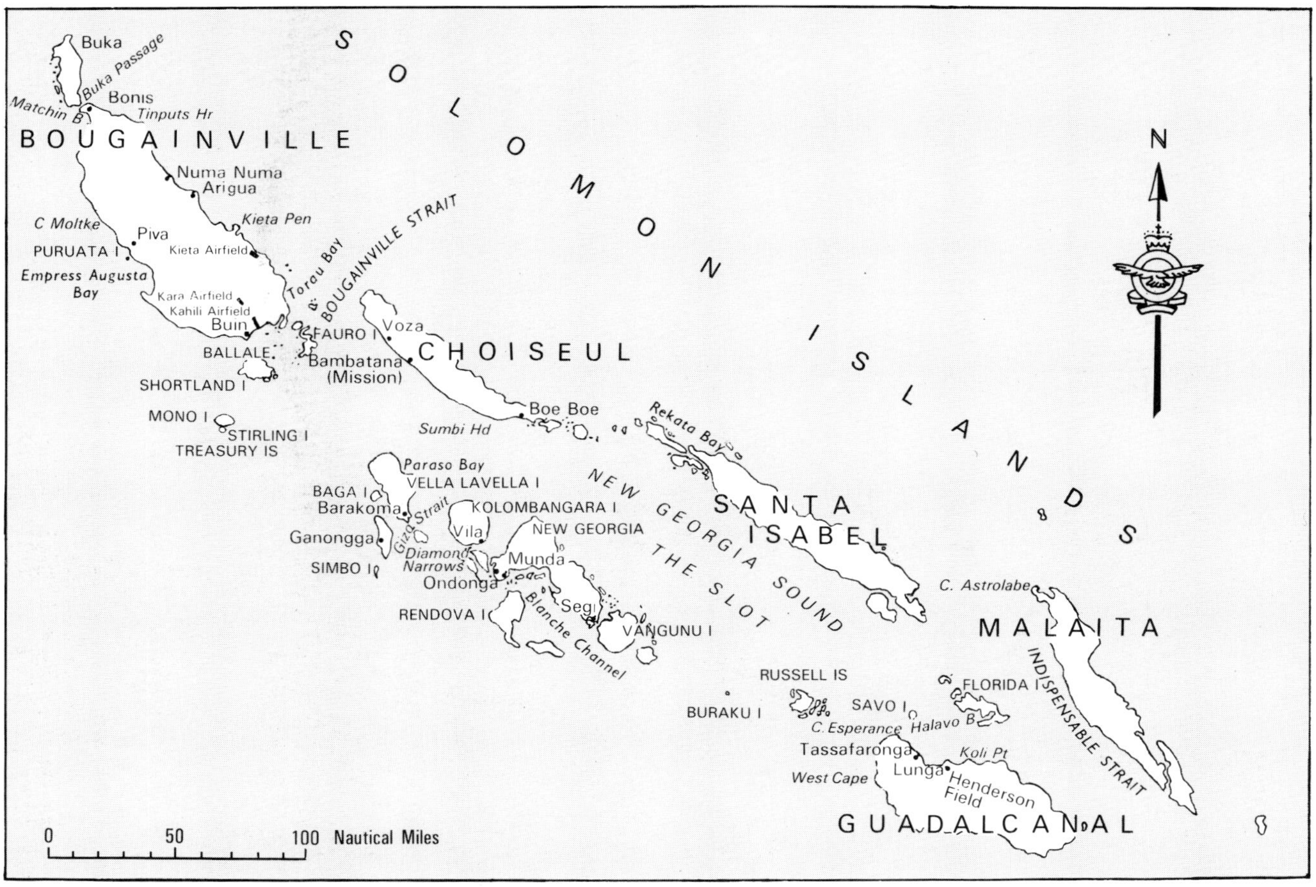

Squadron Leader J. A. Oldfield.

Squadron Leader T. J. MacLean de Lange.

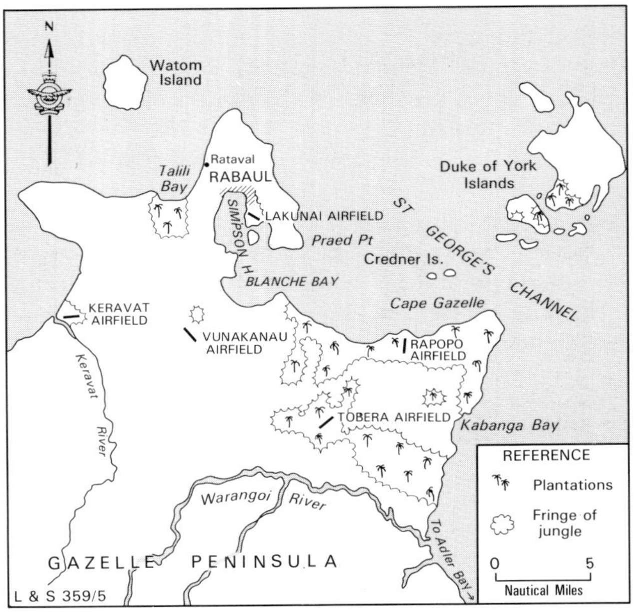

Another New Zealander was shot down but was rescued after six hours in the water.

RNZAF fighter pilots continued to make a name for themselves throughout the Solomons campaign, taking part in every major attack launched against Rabaul. They were popular with American bomber crews because of the way they maintained close cover, resisting the temptation to be drawn into combat elsewhere. The Americans often asked specifically for them for this task.

Squadron Leader Oldfield was promoted to wing commander and took command of the wing. He had the distinction of shooting down the ninety-ninth — and last — aircraft for the Fighter Wing, on 13 February 1944. He was succeeded by Wing Commander S. G. Quill.

It could fairly be said that, by February 1944, the enemy air forces in the South-West Pacific had been defeated, for during that month most of the remaining Japanese aircraft were withdrawn to Truk. Five fighter squadrons of the RNZAF had shared fully in inflicting this defeat. Those that came later — Nos 19 to 26 inclusive — were employed in patrol work and in bombing and strafing ground targets.

Since the start of the Rabaul offensive No. 1 (Bomber Reconnaissance) Squadron, RNZAF, led by Squadron Leader H. C. Walker, AFC, had been stationed at Guadalcanal, with a detached flight at Munda, which the Allies had retaken from the Japanese on 5 August 1943. The squadron was equipped with the first of the RNZAF's new Lockheed PV-1 Venturas, similar to Hudsons but larger and with greater offensive power. The Venturas began

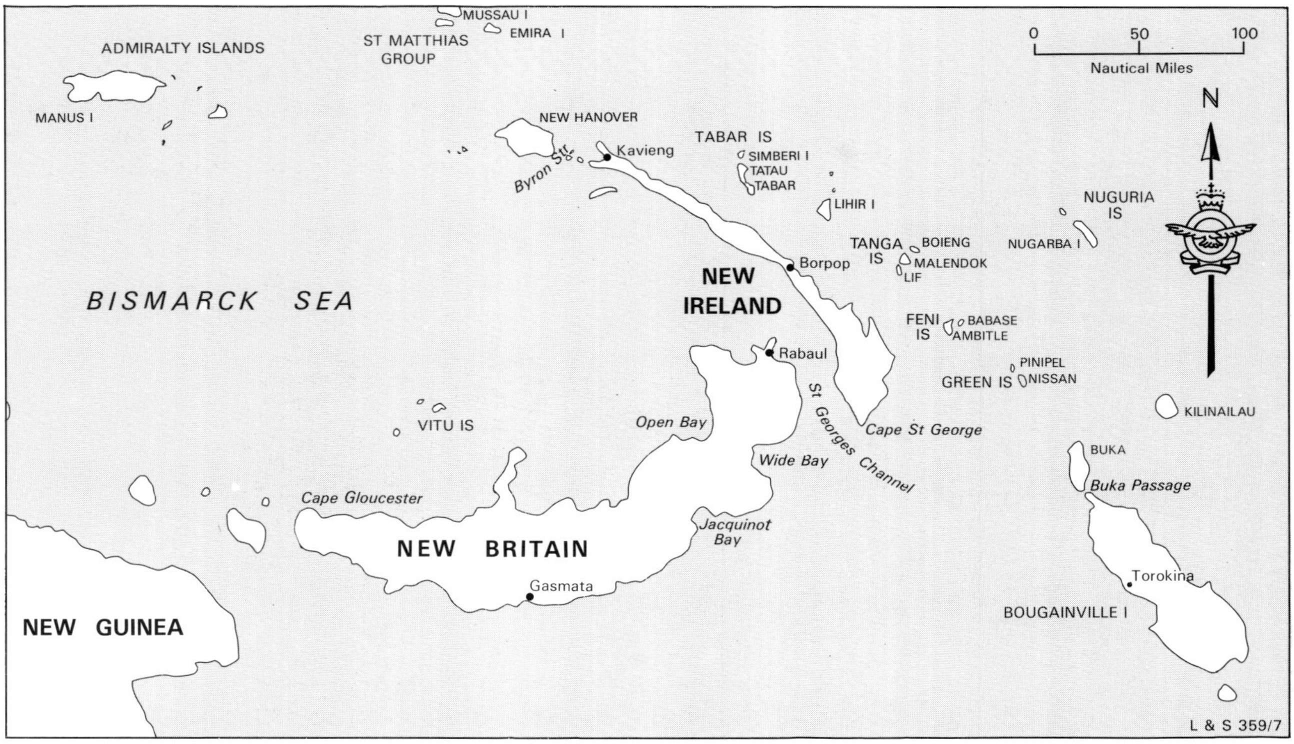

Venturas on a bombing raid in the Northern Solomons, 1944.

operations from Guadalcanal with standard reconnaissance patrols but before long the squadron, particularly the advance flight of six aircraft at Munda, was engaged in bomber strikes and night heckling raids over Rabaul and strafing attacks against Japanese barges along Empress Augusta Bay, Bougainville. The nights of 16 and 17 November were spent mining Buka Passage. With US B-24 Liberator and B-25 Mitchell bombers and Marine Dauntlesses making almost daily attacks on Rabaul, Venturas were detailed to follow the striking forces and search for pilots who had been shot down or had baled out over the sea. (They carried spare inflatable dinghies which could be dropped to airmen in the water if necessary.) Usually the Ventura pilot would call in a Catalina flying-boat, and if possible would orbit the area until it arrived.

Many Allied airmen owed their lives to the fact that they had been seen by members of No. 1 Squadron and their successors. A notable feat concerned two RNZAF Venturas which were attacked by Zeros while on survivor patrols near Rabaul. Piloted by Flying Officer R. J. Alford and Flight Lieutenant D. F. Ayson, the two aircraft had followed the big raid on Christmas Eve 1943. Just as Alford saw a pilot in a dinghy down in St George's Channel, he was attacked by three Zeros. Although his aircraft was hit he shook off the enemy, damaged two of the Zeros and escaped through cloud after signalling the position of the downed airman.

When Ayson's aircraft, now alone over the mouth of the channel, was attacked by a large force of Zeros, the situation looked grim. Flying Officer S. P. Aldridge, the wireless operator/air gunner who was acting as fire controller, was wounded at the height of the action and Warrant Officer W. N. Williams, DFM, the navigator, took over. Despite the heavy odds the Ventura crew shot down three of the Japanese aircraft, possibly destroyed two others and then flew back to base an aircraft that was very extensively damaged. Later they received a personal message of congratulations from General R. J. Mitchell, Commander Air, Northern Solomons.

The first awards made to New Zealand Ventura crew members in the Pacific War followed. Ayson was awarded the DFC; Flight Sergeant G. E. Hannah, who had operated the turret guns and was credited with two and a half enemy aircraft destroyed, was given the DFM, and Warrant Officer Williams received the DFC to add to his earlier DFM.

When No. 1 Squadron returned to New Zealand in mid February 1944, its place was taken by No. 2 (BR)

The primitive living conditions of RNZAF air and ground crews at Guadalcanal are apparent in this wartime photograph. *P. D. Gifford Collection*

An Avenger over Rabaul.

Squadron, under Squadron Leader A. B. Greenaway.

Rabaul and its airfields were the principal targets for Nos 25 and 30 (Dive-Bomber) Squadrons when they moved up into the combat area in March 1944, to be followed later by No. 31 (Dive-Bomber) Squadron. Another important target was the airfield at Kavieng.

No. 25 Squadron formed up at Seagrove, near Auckland, in July-August 1943 under the command of Squadron Leader T. J. MacLean de Lange. The RNZAF had acquired a number of Dauntless dive-bombers from the US No. 14 Marine Air Group. Powered by one 1200 hp Cyclone radial engine, the Dauntless had a maximum speed of 252 mph and a range, with a 1000 lb bomb load fitted, of 1115 miles. It was specifically designed for steep angle dive-bombing. The squadron worked up on these war-weary aircraft before leaving New Zealand late in 1943 for Espiritu Santo. There, more Dauntlesses were taken over and the final work-up training began. Later, new Dauntless aircraft were obtained with which the squadron would go into action.

No. 30 Squadron was formed at RNZAF Station Gisborne in June 1943 under Squadron Leader R. G. Hartshorn. At first it flew Vincents and Harvards taken over from No. 8 (GR) Squadron, which had been based there since early 1942, but as Grumman TBF Avengers became available the squadron converted on to the new type and also went forward to Espiritu Santo, but taking its aircraft with it.

While both 25 and 30 Squadrons were completing their operational training, news came through of trouble on Bougainville. The many thousands of Japanese on the island had become bolder, concentrating their attack on the area held by the Allied forces. At the beginning of March 1944 the Allied perimeter at Torokina extended for about 4 miles along the coast of Empress Augusta Bay and ran inland from the beach to form a rough semi-circle, with a maximum depth of about 3 miles from the sea. The Americans had developed three airfields there, the original Torokina fighter strip, running parallel with the beach, and fighter and bomber airstrips at Piva, about 2 miles inland.

Late in February American intelligence learned of a Japanese plan to attack the perimeter on 7 March. The enemy proposed driving the Allies back to the beaches where a naval attack would, they believed, annihilate them. During the first week of March preparations were made for the defence of the perimeter and at 6 a.m. on 8 March the Japanese launched the artillery bombardment on Allied positions, maintaining this shelling daily throughout the month. It was in such conditions that the Servicing Units of 25 and 30 Squadrons arrived on Bougainville on 15 March.

The New Zealanders experienced their first taste of direct enemy action as soon as their feet touched Bougainville soil. Arriving in the thick of Japanese artillery fire, they began the unenviable task of digging foxholes and improvising their camp only a few hundred yards from the enemy. For two weeks they did not have a normal night's sleep; on some nights they did not sleep at all.

When the aircrews arrived a camp had been established despite the difficult conditions and the Dauntless crews lost no time in getting their first sortie under way. The first aircraft landed at Piva at noon on 23 March and the squadron carried out its first operation at 5.30 a.m. next day. The rest of the Dauntlesses came in that day and the Avengers of No. 30 Squadron and 19 sorties were flown, mainly bombing attacks on Japanese gun positions.

A Dauntless leaves its Bougainville base at the start of a mission.

The shelling of the New Zealand camp continued, especially when ground crews were starting up the aircraft. The shelling would begin at dawn each day and the aircraft would start up in the midst of it, flying around the Japanese positions to pick up the flashes and pinpoint the guns.

No. 25 SU's task was clearcut: to supply the squadron commander each day with 12 serviceable aircraft out of 15. Over almost three months of active operations the SU produced its 12 aircraft quota on all days but one.

But the two squadrons had not been brought forward merely to secure the perimeter. On 26 March both units were detailed to take part, with American squadrons, in an attack on Kavieng airfield. American troops had landed on Emirau Island, 75 miles away, some days before and the Kavieng raid was designed to render the airfield unserviceable and prevent Japanese air reinforcements being flown in from the north. No. 25 Squadron missed the rendezvous but the seven aircraft of No. 30 Squadron, together with 52 American SBDs and TBFs, successfully bombed the target.

After the Japanese counter-attack had petered out the squadrons were used against supply dumps, barge hideouts, bridges and artillery emplacements. They also worked closely with American ground forces bombing enemy troop concentrations and destroying roadblocks. But the squadrons' main effort, while stationed in the combat area, was against Rabaul and its airfields. In conjunction with United States squadrons, they sent over flights of six to 12 aircraft each on most days; the Vunakanau and Tobera airfields, which the Japanese frequently repaired, were attacked time and again. By the end of April the continuous attacks were having an appreciable effect on the AA defences. Although they were to be a formidable hazard for many months to come they were not as strong as they had been earlier in the year.

In the latter part of May 1944 Nos 25 and 30 Squadrons were withdrawn to New Zealand on completion of their tour of duty and disbanded in June. A second Avenger squadron, No. 31, moved to the forward area under the command of Squadron Leader M. Wilkes. (It had formed at Gisborne on the departure of No. 30 Squadron.) Six of its aircraft were flown up and the remainder of the aircrew travelled by 40 Squadron, taking over aircraft at Bougainville supplied by No. 30 SU.

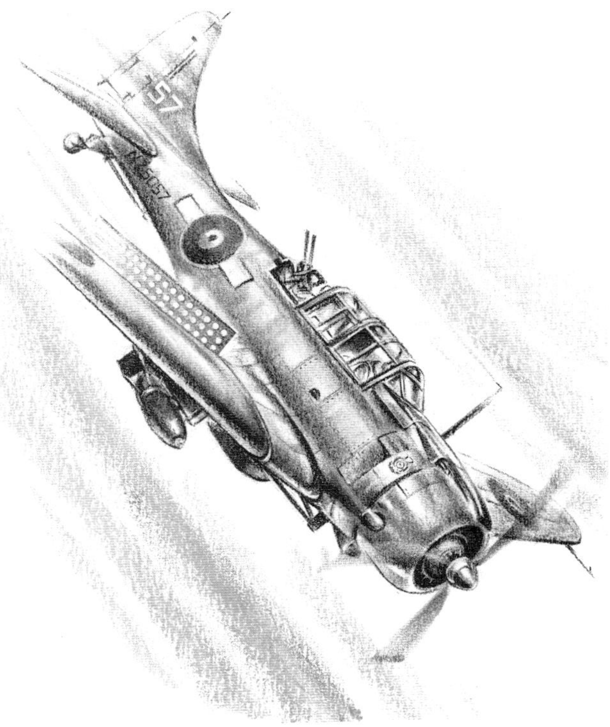

A Dauntless dive-bomber of No. 25 Squadron.

In its two-month tour of duty No. 31 Squadron was engaged in the same types of operations as 30 Squadron, striking against targets at Rabaul and on Bougainville. Its biggest day was 30 June when nine aircraft made three sorties each and dropped 27 tons of bombs on the village of Pikei on the southern shore of Empress Augusta Bay where concentrations of enemy troops were reported.

An Avenger of No. 30 Squadron being bombed up at Bougainville in March 1944. The active volcano, Mount Bagana, is visible in the background.

Crews of No. 25 (Dive-Bomber) Squadron being debriefed after an operation.

During May a new type of target came under attack. The sea and air blockade was practically complete and to eke out their supplies of food the Japanese on Bougainville were growing their own vegetables. The destruction of crops therefore became an important objective as the Allied policy of bypassing isolated enemy garrisons included starving them out. No. 31 Squadron's Avengers were now taking part in the diesel spraying of gardens from the air. These operations were not as easy as they might sound; the spraying had to be done at low level and the aircraft came under very accurate fire from AA gun positions and from small arms.

The squadron flew its last strike on 23 July and with its return to New Zealand and disbandment the era of RNZAF dive-bomber squadrons was over.

By the end of 1943 Nos 16, 17 and 18 (Fighter) Squadrons had all experienced their first combat operations in the Solomons campaign. In 1944 they were followed by Nos 19 and 20 (Fighter) Squadrons. About this time the RNZAF fighter units were re-equipping with Chance Vought F4U Corsairs which were faster and more powerful than the P-40s. The Corsairs were shipped to the New Hebrides, assembled at the RNZAF Base Depot at Pallikulo, on Espiritu Santo, and brought on charge between March 1944 and February 1945. Others were assembled at Hobsonville between May and August 1944 and retained in New Zealand for training purposes. No. 20 Squadron was the first unit to enter the combat area with the new aircraft, arriving on Bougainville on 14 May 1944.

Corsairs of No. 25 (Fighter) Squadron in a revetment at Los Negros, February 1945.

CHAPTER TWELVE

'Dumbo'

By early 1944 Japanese air opposition was no longer a major threat in the South-West Pacific and by the end of that year it was almost non-existent. Almost, but not quite, for as late as 29 April 1945 two Japanese aircraft flew over Los Negros about midnight and dropped bombs on the floating dock in Manus Harbour, a few hundred yards from the RNZAF sleeping quarters. But although the skies were safer for Allied airmen, there were other hazards, not the least of which was the deadly accuracy of the Japanese anti-aircraft guns at Rabaul and Kavieng. Enemy gunfire from these fortifications caused the destruction of many Allied aircraft and when pilots and crews parachuted or crash landed in the sea they did so in the comforting knowledge that 'Dumbo', American or New Zealand, was never far away.

'Dumbo' was the affectionate title bestowed by Allied airmen on the morale-boosting air-sea rescue operations of patrol squadrons which, at this period of the war, flew Catalinas. The name derived from the big-eared, benevolent flying elephant of a popular Walt Disney cartoon film. Cartoon Dumbos were painted on the fuselages of the flying-boats, one for each rescue effected.

The RNZAF had two flying-boat squadrons in the Pacific area, Nos 5 and 6. They flew long-range anti-submarine patrols, shipping escorts and air-sea rescue missions. Although submarines were reported and occasionally sighted, the regular ocean patrols were mostly routine and largely uneventful. The rescue missions, by contrast, were generally successful and as they were often flown in hazardous conditions they brought a measure of excitement to the operations and were satisfying and morale boosting to the crews.

No. 6 (Flying-Boat) Squadron was the first of the two to be equipped with Catalinas and move into the forward area. It was formed in Fiji on 25 May 1943 under Hastings-born Wing Commander G. G. Stead, DFC, RAF, and it absorbed the Singapore Flight of No. 5

A Catalina flying-boat of No. 6 Squadron, at Hobsonville, on arrival from Lauthala Bay in May 1943. This was the first RNZAF Catalina to land in New Zealand waters. Standing in the cockpit are, from left: Squadron Leader R. B. L. MacGregor, a flight commander, and Wing Commander G. G. Stead, RAF, the squadron commander, and standing behind them are Flight Lieutenant K. H. Day, an engineer officer, and Flying Officer T. D. Brewer. The other crew members are not identified.

A 'Dumbo' Catalina.

Squadron after that unit was disbanded in November 1942. (Between November and May the Singapore Flight was loosely and unofficially known in some quarters as No. 6 Squadron.)

The first Catalina was delivered to Lauthala Bay, Fiji, on 3 April 1943 by an American naval crew who had flown it from San Diego on the west coast of North America. Subsequent aircraft were brought to New Zealand by RNZAF crews who had been attached to the US Pacific Ferry Command for this purpose and for prior training in America.

Wing Commander Stead had an establishment of 24 aircraft, 40 aircrew and 111 ground crew. One flight commander, Squadron Leader R. B. L. MacGregor, and several crews had earlier been attached to the operational headquarters of COMAIRSOPAC in Segond Channel to study United States naval procedures in general and Catalina operations in particular, and this experience proved invaluable during No. 6 Squadron's intensive flying training programme.

One of the first sea rescues by an RNZAF Catalina predated the official formation of No. 6 Squadron and was carried out by the crew of Catalina NZ4001 captained by Squadron Leader MacGregor. This was on 2 May 1943 when nine survivors* of the torpedoed American ship *Vanderbilt* were rescued 120 miles south of Suva. The Catalina found the life-raft containing the men and circled overhead for three hours, waiting in vain for rescue ships to reach the scene. Squadron Leader MacGregor then landed successfully, despite the very rough seas which damaged the hull, and the rescue and subsequent take-off were accomplished without further incident. On 1 June MacGregor was awarded the Air Force Cross for this mission. (Four days later he and his crew were posted as missing, believed killed, on operations.)

On 1 October 1943 the main body of the squadron, now under the command of Wing Commander I. A. Scott, left Fiji for Segond Channel, where they came under the operational control of the US Navy. Shortly before Christmas the squadron moved forward to Halavo Bay, on Florida Island, and the first operational flight from Halavo was flown on Christmas Day.

The first New Zealand 'Dumbo' mission from Halavo Bay was flown on 26 January 1944 when an aircraft captained by Flying Officer W. B. Mackley, DFC, rescued 10 crew members of a United States B-24 Liberator bomber which had been ditched in the open sea 200 miles to the north. Open sea landings such as this were always hazardous except in the most ideal weather and sea conditions. Strong winds and heavy swells could (and sometimes did) damage the flying-boats on landing and make it impossible for them to take off again.

On 4 February 1944 a Catalina captained by Flying Officer D. S. Beauchamp, on a routine patrol 100 miles south of Nauru, discovered five men in rubber dinghies, survivors from an American Liberator forced down while returning from a raid on Kwajalein. The men had been adrift for six days and nights. The Catalina's hull was severely strained in the landing, rivets popped and the flying-boat began to take in water. It was Beauchamp's first open sea landing and he made several experimental manoeuvres before electing to take off parallel with the heavy Pacific swell. He received an immediate award of the DFC for this operation.

Another notable 'Dumbo' that year was the rescue of an American B-25 Mitchell crew under the very noses of the Japanese at Rabaul. The bomber was shot down by shore-based anti-aircraft guns on 9 March 1944 and landed in the sea about 5 miles offshore, 20 miles south of Cape St George. The crew managed to launch two life-rafts and climb into them before the bomber sank. Flying

An air-sea rescue by an RNZAF Catalina.

* Some records give the number as eight.

Officer Beauchamp, who was on patrol in a Catalina of No. 6 Squadron about an hour's flying time away, responded to the call and two RNZAF Venturas circled the downed airmen until the Catalina arrived. The sea was smooth but the hazards this time came from the heavily fortified coastline. Flying Officer Beauchamp put the Catalina down and taxied up to the dinghies. The Americans were rescued after being in the sea for only 55 minutes.

No. 5 (Flying-Boat) Squadron was reactivated in July 1944, equipped with Catalinas, under the command of Wing Commander J. W. H. Bray. After the crews had completed operational training at Lauthala Bay, the squadron moved to Segond Channel where it flew anti-submarine patrols, shipping escorts and rescue missions throughout the rest of the war, with a detached flight operating from Funafuti.

RNZAF aircrew also had cause to be grateful to American 'Dumbo' Catalinas. For example, on Christmas Day 1944 Squadron Leader J. R. C. Kilian, commanding officer of No. 19 (Fighter) Squadron stationed on Emirau Island, encountered anti-aircraft fire over Kavieng, which crippled his Corsair. He parachuted into the sea from 1500 feet some 40 miles from Emirau and managed to inflate his rubber dinghy and climb in. Flight Sergeant G. J. Ronayne, who was flying as his No. 2, remained overhead circling and advised headquarters at Emirau of the situation. An American air-sea rescue Catalina was alerted and less than 45 minutes after he had baled out Kilian was on his way back to Emirau.

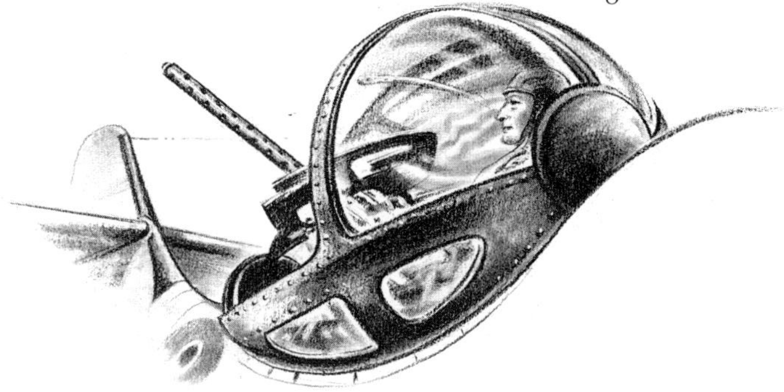

The distinctive Catalina gun 'blister'.

Flying Officer D. S. Beauchamp, DFC, from a portrait by Russell Clark, circa 1943.

This was the second rescue of a New Zealand pilot within 24 hours in which American 'Dumbo' took part. On Christmas Eve Pilot Officer K. W. Wright, also of No. 19 Squadron, was forced to land in the sea 25 miles from base after his Corsair was struck by enemy fire during a raid on New Ireland. Emerging from a dense cloudbank, Flying Officer E. T. Aickin noticed that Wright was no longer flying alongside him. He turned back and flew low over the sea, searching a wide area. Oil patches on the water and brilliant yellow dye guided him to the spot where Wright was drifting in his dinghy. Emirau was advised and both 'Dumbo' and a patrol boat were dispatched.

A Catalina of No. 3 (Flying-Boat) Operational Training Unit, Lauthala Bay, becomes airborne. Catalinas equipped this unit as well as Nos. 5 and 6 Squadrons.

Catalina NZ4038 of No. 5 Squadron at Segond Channel, Espiritu Santo. The photograph shows the detachable beaching gear being dragged ashore after launching.

The American Catalina was quickly overhead but a heavy sea and reduced visibility made an ocean landing impossible. Together the flying-boat and the fighter circled the dinghy while rain squalls obscured the area. The hours passed. Wright was lost from view and the weather deteriorated further. Aickin's Corsair was now low on fuel and he turned reluctantly for base, to be relieved by another aircraft. Then the weather cleared and the Catalina relocated Wright and dropped a flare to guide the patrol boat to the spot. When the vessel was alongside, the 'Dumbo' crew had the pleasure of seeing the completion of a rescue operation they had begun.

These two rescues were perhaps typical of the way in which 'Dumbo' operated. Certainly the protection afforded by 'Dumbo' was enormously appreciated by bomber crews and fighter pilots alike who knew the hazards facing the flying-boat crews and respected both their special skills and their courage and devotion to duty. In all, Nos 5 and 6 Squadrons collectively rescued at least 150 Allied servicemen from the sea.

RNZAF Marine Section puts a high-speed launch through its paces.

CHAPTER THIRTEEN

Final Stages of the Pacific Campaign

By late 1944 the South Pacific no longer dominated the world's headlines as the armed might of the United States swept the war closer to Japan's own shores. Admiral Halsey, who had directed the operations which carried the Allies from Guadalcanal to the Bismarcks, handed over command in the South Pacific to Vice Admiral J. H. Newton and, as Commander of the American Third Fleet, turned his attention to the Carolines and the Philippines. US carriers struck at bases in the Philippines in September 1944, inflicting heavy losses on the enemy, and the following month they attacked Formosa.

General MacArthur's South-West Pacific forces, with New Guinea retaken, had by then bypassed a long chain of islands — including the Solomons — and were ready to return to the Philippines. The invasion was launched in mid October 1944 with the support of fast aircraft carriers, escort carriers and the USAAF's 5th and 13th Air Forces to the south. The Japanese Navy provided deadly opposition and the first kamikaze suicide attacks were launched against the carriers. There were heavy casualties on both sides but eventually the tables were turned.

By early 1945 the British Pacific Fleet was operating successfully with United States naval forces in waters close to Japan's shores and US forces had landed first on Iwo Jima and then on Okinawa.

On 6 August the first atomic bomb was dropped on Hiroshima from a US B-29 Superfortress bomber. Two days later the Soviet Union attacked Japanese forces based in Manchuria. On 9 August a second atomic bomb was dropped on Japan, this time over the city of Nagasaki. Crushed and bewildered, Japan agreed on 14 August 1945 to unconditional surrender.

In the last 11 months of the South-West Pacific campaign, RNZAF units became part of a New Zealand Air Task Force (ZEAIRTAF), which had a specific role in the area — the conduct of air operations in the Northern Solomons. Previously individual New Zealand units had operated within American commands, but the Americans were intending to move their 1st Marine Air Wing to the Philippines in September 1944 and it was agreed that, when they moved out, a New Zealand Air Task Force should assume operational control.

ZEAIRTAF was officially formed on 1 September and Group Captain G. N. Roberts arrived on 16 September to take command. (He was shortly to be promoted to air commodore.) The headquarters was established at Bougainville, and subordinate formations, called field headquarters, were set up at Piva (Bougainville), Green Island, Emirau Island (in the St Mathias Group) and Los Negros (in the Admiralty Islands) where New Zealand squadrons were based. Delays affected the 1st Marine Air Wing, however, and it did not quit the area until July 1945. It had been intended that the commander, New Zealand Air Task Force (COMZEAIRTAF), should take over from COMAIRNORSOLS on 1 November 1944 but, because of the delays in transferring the Americans, the war was almost over by the time the Northern Solomons became a New Zealand command.

Pacific base control tower.

The operations room at RNZAF Field Headquarters, Emirau, late in 1944.

It was not a particularly rewarding period for the RNZAF in the Pacific. Japanese air opposition was virtually non-existent, and the Corsair squadrons were operating as fighter-bombers only, with none of the close combat which had stimulated the pilots of the earlier squadrons. As far back as April 1944 the Joint Chiefs of Staff Committee in Washington had proposed that the future employment of the RNZAF in the South-West Pacific be restricted to a handful of bomber reconnaissance and flying-boat squadrons on garrison duty only, which would have meant the scrapping of all New Zealand's fighter squadrons. The RNZAF was quite opposed to taking a back seat at this stage.

An immense amount of money and effort had gone into the training and equipping of the Corsair squadrons. At a national level, if New Zealand dropped out of the Pacific War before it ended she would have had small reason for making her voice heard in subsequent peace councils. In Washington there was considerable discussion on the future role and location of New Zealand's air units and many alternative proposals were debated. The American COMSOPAC and his subordinate commanders knew well, however, the New Zealanders' fighting qualities and airmanship standards and were keen to retain them in spite of the attitude of their superiors in the Pacific and in Washington. Eventually it was agreed that the New Zealand units in the area should be included in the Northern Solomons Allied forces that were to be handed over to General MacArthur.

In the last phases of the conflict in the Northern Solomons, RNZAF fighter and bomber squadrons mounted an offensive calculated to soften up the Japanese for an attack by ground forces which would break their resistance completely. The campaign was designed to render their anti-aircraft artillery ineffective, destroy their airfields and ammunition dumps, disrupt their transport systems, reduce their gardens and weaken their morale. In these tasks, operating from Bougainville, Green Island, Emirau, Los Negros and finally Jacquinot Bay, the New Zealand squadrons and servicing units maintained high standards of performance.

At home the last 12 months of the war brought countless organisational changes within the RNZAF — stations closed down, units were disbanded, functions were transferred. One development which was of major significance was the growth of an air transport organisation which would play a vital role in the RNZAF over the next 43 years.

Obviously the strength of the RNZAF in the forward area and the lengthening of its lines of communication made more transport squadrons essential. By August of 1944 No. 40 Squadron had an establishment of 37 aircraft — 16 Dakotas, nine Lodestars and 12 Hudsons. (Made redundant as bomber reconnaissance aircraft by the introduction of Venturas, some Hudsons had been converted to troop carriers late in 1943.) It was now timely to divide the fleet and create a new long-range transport squadron. No. 41 Squadron was formed at Whenuapai on 15 August under the command of Wing Commander H. C. Walker, AFC,* taking over the Lodestar and Hudson aircraft of No. 40 Squadron. These were soon replaced by 20 new Dakotas which came to hand in the first six months of 1945.

During 1945 the Air Force's transport services were augmented by a flight of four Short Sunderland III flying-

* Squadron Leader A. F. H. Tye, DFC, led the unit until Wing Commander Walker arrived in September.

Four well-known officers who commanded RNZAF bases in the Pacific during 1944–45. From left: Wing Commanders P. A. Matheson, Field Headquarters Green Island; F. R. Dix, Field Headquarters, Emirau Island; M. Wilkes, RNZAF Station Guadalcanal and H. R. Wigley, Field HQ, Piva, Bougainville. Squadron Leader W. B. Cowan, who is not in the photograph, commanded the other Field HQ at Los Negros.

Sunderland NZ4102, named *Tokomaru*, was one of four Mark III Sunderlands which entered service with the RNZAF as transport aircraft in December 1944.

boats,* which had been delivered by ferry flight from the United Kingdom in December 1944. The aircraft were based at Mechanics Bay, Auckland, from where they made regular flights to Fiji, New Caledonia and Espiritu Santo. They had the advantage of being able to carry twice the payload of a Dakota.

The other transport unit to achieve squadron status — and a lasting niche in the postwar Air Force — was the Communications Flight which had operated air transport services out of Rongotai (Wellington) almost daily throughout the war, equipped at various times with Harvards, Oxfords, Dominies and a variety of light ex-civilian aircraft. Renamed No. 42 (Transport) Squadron on 20 December 1943, still under Wing Commander J. D. Hewett, AFC, it was kept very busy during 1945, shouldering more of the internal transport burden while No. 41 Squadron shared the overseas tasks with No. 40 Squadron.

When the Japanese finally surrendered, all RNZAF fighter squadrons, with the exception of those resting and reforming in New Zealand, were operating in the Solomons-Bismarcks area. Four, with their servicing units, were on Bougainville, two were at Los Negros and three were at Jacquinot Bay. A move forward to Borneo had been planned but had not taken place. Of the bomber reconnaissance squadrons, four had been disbanded as there was little prospect of further employment. The other two, Nos 2 and 4, were at Jacquinot Bay and Los Negros respectively. The two flying-boat squadrons were at Espiritu Santo and Halavo and there was a flying-boat OTU at Lauthala Bay.

Squadron Leader J. J. de Willimoff, who commanded No. 23 (Fighter) Squadron at Emirau and Los Negros in 1944 and 1945. This photograph was taken when he was Director of Operations and Flying Training at Air Force Headquarters, Wellington, after the war.

Headquarters of the New Zealand Task Force was at Bougainville. Of the subsidiary field headquarters which had been formed the previous year, those at Piva and Los Negros were still operating. The unit on Green Island had moved to Jacquinot Bay and the Emirau headquarters was in the process of moving to Los Negros. No. 1 (Islands) Group Headquarters had closed down on 1 August and RNZAF Station Guadalcanal was responsible for the

* In RNZAF service, each of the four Sunderlands of the Flying-Boat Transport Service was named after one of the canoes in the Maori migration to New Zealand — Tainui, Tokomaru, Mataatua and Takitimu.

Watched by General Douglas MacArthur (left), Air Vice-Marshal L. M. Isitt, Chief of the Air Staff, signs the Japanese surrender document aboard USS *Missouri* in Tokyo Bay on 2 September 1945.

administration of units within its area. No. 1 (Islands) Works Squadron still had its headquarters at Guadalcanal, with detached elements at Espiritu Santo and Bougainville. No. 2 (Islands) Works Squadron was at Los Negros, where it had moved earlier in 1945, and a detached flight was at Jacquinot Bay.

Most RNZAF units heard of the Japanese acceptance of the Allies' surrender terms with profound relief. The reaction at Camp Waitemata on Bougainville was no doubt typical of RNZAF units in the Pacific. The rejoicing began on the evening of 10 August, with rumours (which were true) that Japan had made the Allies an offer of surrender. Before long the camp rang with the sounds of celebration. The sky was illuminated every few seconds as Allied guns released jubilant salvoes into the air while tracer and machine-gun fire added to the fireworks. Down at the harbour ships' sirens and bells were sounding and in the darkness someone was playing the bagpipes. Hundreds of motor vehicles were sounding their horns while rockets, parachute flares and Very lights were streaking the sky with light and colour. Thousands of United States, Australian and New Zealand servicemen had chosen not to wait for the big powers' reply to Japan's offer and had decided 'this is it'.

On 12 August the 2nd Australian Corps completed arrangements to achieve the full surrender of the Japanese forces on Bougainville, should Tokyo accept the Allies' terms. All RNZAF air strikes were cancelled and an RAAF Wirraway dropped 18,000 leaflets telling the enemy that in the event of Japan surrendering all Japanese officers should instruct their men to cease fire.

Four days later five RAAF Beauforts skimmed over Bougainville at low level, carrying the news of Japan's surrender to all enemy ground forces in boldly painted Japanese symbols on the undersurfaces of the mainplanes. At the same time RAAF Wirraways and Boomerangs flew low over Kara and Kahili airstrips, risking fire from anti-aircraft batteries, to drop messages which, the accompanying pamphlets instructed, were to be taken to Lieutenant General Masetane Kanda, who commanded the Japanese division.

Group Captain G. N. Roberts. He commanded the New Zealand Air Task Force in the Pacific in the final stages of the Pacific War, as an air commodore.

Although operations ceased on 15 August the formal surrender ceremonies did not take place until the following month. At the principal ceremony on board USS *Missouri* in Tokyo Bay on 2 September 1945, Air Vice-Marshal Isitt represented New Zealand. In the South-West Pacific Air Commodore G. N. Roberts, AOC New Zealand Air Task Force, was the senior New Zealand officer present when on 6 September Lieutenant-General Imamura and Vice-Admiral Kusaka, commanding the Japanese forces in the area, surrendered to Lieutenant General V. A. H. Sturdee, GOC 1st Australian Army, aboard the British aircraft carrier, HMS *Glory*, off Rabaul.

Two days later at Torokina, Air Commodore Roberts represented New Zealand again when General Savige received the formal surrender of Lieutenant-General Kanda and his forces on Bougainville.

With the war's end the RNZAF was faced with three immediate tasks — returning to New Zealand the 7000 men stationed at bases in the South and South-West Pacific, repatriation of New Zealand prisoners of war in Malaya and the demobilisation of men and women who would not be required for further service. Nos 40 and 41

The flight deck of a Sunderland flying-boat.

Squadrons between them transported most of the RNZAF personnel from the Pacific, but about 2000 were carried in the Union Steam Ship Company's inter-island vessel, *Wahine*, which made three trips under charter. Catalinas of No. 5 Squadron — No. 6 had by then been disbanded — and the four Sunderlands also assisted.

The repatriation of New Zealand prisoners of war from the Far East was handled by the air transport organisation at Whenuapai, with Squadron Leader M. L. Pirie in charge of transport arrangements. Dakotas were fitted out as air ambulances, medical staffs and supplies of food, clothing and comforts were assembled, and a ground component was organised to inspect and, if necessary, service the aircraft at staging points along the route.

Preparation was completed within a few days of the Japanese surrender but the aircraft could not leave for Malaya until the relieving British forces were about to enter Singapore. The first two aircraft left New Zealand on 4 September and the British forces arrived at Singapore the following day. Both the New Zealand transport aircraft met with disaster en route when a United States Liberator taxied into them at Morotai Island, north-west of New Guinea, destroying one and damaging the other. Replacement aircraft were speedily deployed by Whenuapai.

The first two RNZAF aircraft arrived at Kallang airfield, Singapore, on 13 September. Pirie acquired a house nearby and set up a headquarters there. He was joined by Wing Commander T. J. MacLean de Lange, who had been the RNZAF Liaison Officer in India since 25 Squadron's disbandment and who had been assigned to work with the British RAPWI* organisation and locate New Zealand prisoners of war in the area.

It was quickly discovered that RAPWI had little or no information on the whereabouts of New Zealanders and the RNZAF detachment had to do its own research. The task was completed by the middle of October. Altogether the detachment repatriated 156 New Zealand prisoners of war and civilian internees and two Australians.

During the war Nos 40 and 41 Squadrons between them carried 37,000 passengers into or out of Whenuapai, besides nearly 4,000,000 lb of freight and 1,500,000 lb of mail.

In the last four months of 1945 the Air Force's principal task was returning personnel to civilian life as quickly as possible. A total of 1642 men and women were released in August, and in the next four months the figures were 8215, 3503, 4328 and 3268 respectively. The year ended with the strength of the RNZAF at 12,228,9899 in New Zealand and the remainder either in the Pacific or in the United Kingdom, awaiting repatriation.

RNZAF involvement in the Pacific Campaign reached its peak in February 1945 when there were 7929 personnel in the Islands. This figure had been reduced to 7276 at war's end.

RNZAF casualties in the Pacific War included 82 killed on active service, 256 presumed dead, seven deaths from natural causes, 58 seriously wounded or injured and four prisoners of war.

* Repatriation of Allied Prisoners of War and Internees.

CHAPTER FOURTEEN

With the Occupation Forces in Japan

When the war against Japan came to its dramatic close the Royal New Zealand Air Force immediately found itself committed to a new responsibility — contributing a war-seasoned fighter unit to the British Forces of Occupation in Japan. Although 13 New Zealand fighter squadrons had served in the South-West Pacific campaign, it was No. 14 Squadron that was selected, gaining a lasting niche in the peacetime air force after its sister units were honourably disbanded.

If 14 Squadron was regarded as a logical choice of unit, so too was the choice of its commander. Squadron Leader J. J. de Willimoff, DFC, had been an original member of 15 Squadron, he had completed five separate tours of duty in the Pacific and he had commanded No. 23 (Fighter) Squadron in that theatre in 1944–45. With a proven operational record, a strong personality and a sense of fun that infected others, he was the right sort of leader for the Occupation Force unit.

De Willimoff had as his flight commanders Flight Lieutenants L. A. Rayner, DFC, and D. M. Hutton. Flight Lieutenant W. Davenport Brown, DFC, a former gunnery leader with No. 75 (New Zealand) Bomber Squadron, was the squadron adjutant, and Flight Lieutenant C. T. B. Fitzgerald, MBE, was the engineer officer. The 250 men who comprised the squadron paraded before the Chief of the Air Staff, Air Vice-Marshal Sir Leonard Isitt, at RNZAF Station Ardmore on Friday 15 February 1946.

The men had been training in some of the hottest weather of the driest summer in years. In charge of their ground training had been Squadron Leader J. E. Duncan, MBE, the Air Force's chief disciplinarian, who was destined to exert his unique influence for some years afterwards from a desk at Air Force Headquarters. As Warrant Officer Duncan, he was known throughout the Service as 'The Bull' because of his stentorian voice which had a remarkable range. His nickname continued to be used throughout the postwar years. He was almost certainly the best-known member of the RNZAF in New Zealand.

Two hours before the formal parade, Squadron Leader R. M. McKay treated the assembly to a magnificent display of jet aerobatics, the first ever seen in New Zealand. He was flying the country's first jet aircraft, a Gloster Meteor*, in which he was to astound and delight New Zealanders everywhere in the months ahead in the course of a Dominion-wide demonstration tour.

On 8 March 1946, after a farewell parade down

* The Gloster Meteor was the first jet-propelled aircraft to enter RAF service. In 1945 the Air Ministry decided to lend Meteors to Canada, Australia and New Zealand, for indoctrination purposes. The loan was seen as a gesture of appreciation for those countries' assistance in the experimental development of the jet engine. (Group Captain G. E. Watt was a noted New Zealander closely associated with it.) Squadron Leader R. M. McKay, an RNZAF officer who was on the staff of the Empire

Squadron Leader J. E. Duncan, MBE, the RNZAF's chief disciplinarian, known to all as The Bull.

Squadron Leader R. M. McKay and Gloster Meteor NZ6001, the RNZAF's first jet aircraft, at Ardmore, 15 February 1946.

Auckland's Queen Street, the men of No. 14 (Fighter) Squadron embarked in the British aircraft carrier *Glory*, their 24 Goodyear FG-1D Corsairs neatly parked on the upper deck, together with the unit's motor vehicles. A flight of Corsairs from RNZAF Station Ardmore flew past in salute as *Glory* made passage out of Waitemata Harbour.

The New Zealand contingent reached Kure on 25 March and were taken to the former naval air station at Iwakuni which was to be their new home. Their duties there had a significance they were not slow to appreciate.

One of the highlights of the squadron's sojourn was the period about September when Koreans were displaying a marked reluctance to return to their homeland. They swarmed back to Japan in all sorts of decrepit vessels. Corsair fighters flying dawn and dusk patrols over the sea between Korea and Japan spotted and reported many of these vessels, which were promptly given a surface escort; and each passenger was provided with a return passage to Korea.

Among other useful spotting missions, 14 Squadron found a number of 40 mm anti-aircraft batteries, complete with radar devices, the location of which had been conveniently withheld from the occupation authorities. The squadron also assisted in keeping close watch over the schools, one-time incubators of Japanese militarism. Pilots checked to see that pupils were not being lined up in anything like a military formation. Another routine task was to overfly damaged Japanese airfields to check that the Japanese were not filling in the bomb craters.

Central Flying School at RAF Hullavington in 1945, was assigned to Meteor conversion training while in the UK. A Meteor Mk III, EE395, was shipped to New Zealand in November, arriving on 23 December 1945. It was allotted the New Zealand serial number NZ6001. It made its first flight — the first by a jet aircraft in New Zealand — on 11 February. The loan period was extended and eventually the aircraft became the property of the RNZAF upon payment of a mere £5,000.

A Gloster Meteor in flight.

With RAAF Mustang and RAF Spitfire squadrons, 14 Squadron patrolled an area about 200 miles square of mainly rugged terrain. The original ground and air components were almost completely replaced early in 1947 by new arrivals from New Zealand.

Marshal of the Royal Air Force, Lord Tedder, then Chief of the Air Staff in Britain, addressed this tribute to Squadron Leader de Willimoff late in 1946; 'It has been a great tonic to meet and talk to many of your fellows and to see the high standard of efficiency and initiative they have achieved and the fine spirit with which they have tackled the problems that faced them. As they did in the

No. 14 (Fighter) Squadron on parade, Iwakuni, Japan, March 1946.

war, so in these difficult days of early peace they have set a high standard. I take my hat off to them.'

This reputation lost none of its lustre but was, if anything, further enhanced in the next phase of 14 Squadron's sojourn in Japan under Squadron Leader D. F. St George, who succeeded de Willimoff. The squadron, which was yet to undergo a third change of personnel, carried on its good work from Bofu whence it had moved in February 1948. St George, who had commanded No. 17 (Fighter) Squadron in the Pacific Campaign, was also a popular and efficient squadron commander. In his first year he had Flight Lieutenants G. W. Annand and J. R. Claydon as his flight commanders and when their tour of duty was completed they were succeeded by Flight Lieutenants E. J. Shaw (an early member of No. 485 Squadron) and N. H. Bright. Flight Lieutenant H. A. Farrar was the engineer officer.

The squadron had its share of untoward incidents. Flight Lieutenant G. R. B. Highet had an engine failure on take-off and his Corsair overran the airfield into fairly soft ground, overturning and pinning the pilot underneath. He was imprisoned there for about an hour before the fuselage could be hacked through and he was rescued. There were other emergencies and one casualty, Flight Lieutenant C. W. N. Wright losing his life when his aircraft crashed on take-off.

One of the luckiest men in the squadron was Flying Officer R. C. C. Nairn. He was flying number four in a formation coming in to land, when his Corsair crashed and broke up. Rescue parties found the cockpit section reasonably intact, right side up, with the pilot strapped in and unconscious, while the engine, mainplane and petrol tank were all some distance away. Nairn, uninjured, was on his feet the next day.

By late 1948 the RNZAF was re-equipping with de Havilland Mosquito fighter-bombers and the Air Staff intended that 14 Squadron should become one of the new Mosquito squadrons. (This, in fact, did not eventuate and its new role was to be as a multi-engine conversion unit, initially with Oxfords.) Also, the Corsairs were nearing the end of their operational life. (They were all destined shortly to be reduced to scrap metal.) No. 14 Squadron embarked in the mv *Westralia* on the morning of 25 November 1948 and came home to an air force that was gradually, if slowly, being reshaped for the postwar era.

The Berlin Airlift — unloading supplies at Gatow airport.

CHAPTER FIFTEEN

Germany and Malaya

The end of the Second World War did not spell the end of conflict. The suppressed antagonism between two rival ideologies flared up soon after the instruments of surrender had been signed by Germany and Japan. The rising tension between the interests represented largely by Soviet Russia on the one hand and Great Britain and the United States on the other was labelled the Cold War.

After the 1939–45 war ended, the western half of Berlin was occupied by the Western Powers whose occupation forces were placed well inside the Russian zone of Germany. The Western Powers had right of access to Berlin from their own respective zones but in May 1948 the Russians took the startling step of closing all road and rail links into Berlin from the Allied zones.

In West Berlin was a population of more than two million Germans, together with a few thousand Allied troops. Since the end of the war this city, devastated by aerial bombardment and with its industries virtually destroyed, had only just begun to move along the path to recovery. Coal was required for public utilities, raw materials for West Berlin's industries had been imported from several regions, and the total imports of all kinds for the restoration of the city amounted to about 13,500 tons daily. When the surface routes were arbitrarily closed by the Russian blockade, the only alternative access was by air.

The Western Powers countered the Soviet blockade by mounting an airlift of essential supplies to the beleaguered city. In this massive airlift, codenamed Operation Plainfare, members of the RNZAF played a useful part. Group Captain R. J. Cohen was Senior Air Staff Officer at No. 46 Group, RAF, which controlled the operation, and a detachment of three crews commanded by Flight Lieutenant C. J. Fraser was contributed by No. 41 Squadron, but flying RAF aircraft.

There were three air routes into Berlin from the west, by agreement of the powers concerned. Each 20 miles in width, they converged on Berlin in the shape of an arrow's head. The northern corridor led from Hamburg, the central air lane extended due east from Hanover (both from the British zone) and the southern corridor — by far the longest — stretched from the American zone opposite Frankfurt. There were two airfields at the apex of the arrow — the Berlin airport of Tempelhof, and Gatow, a former Luftwaffe training centre. Near Gatow was the Havel Sea, used by the RAF to operate Sunderland flying-boats in the airlift.

The principal requirement was that the airlift should bring in supplies of food. The planned target for Gatow was to land one aircraft and dispatch another every three minutes; a pattern was determined by the time taken for an aircraft taking off to move from its holding point at the downwind end of the runway, roll and become airborne. In practice this took 90 seconds or less and provided time for the dissipation of the slipstream set up along the runway by the departing aircraft. Then, at the end of this 90-second period, the next landing aircraft was planned to touch down, complete its landing run, turn off the runway and proceed to the unloading area — another 90 seconds.

On 26 June 1948, two days after the Western Powers' land routes to Berlin had been sealed off by the Russians, the airlift began with United States aircraft. Two days later, the first RAF plane, laden with flour, took off from Wunstorf, near Hanover, and flew to Gatow. By July it was estimated that about 4000 tons daily could be flown into Berlin by the combined fleets of British and American aircraft. At that stage no one imagined that this operation was to be anything but a short-term affair, yet what had just begun was the greatest air supply operation ever undertaken.

Control was anything but simple. At the height of the operation there were as many as 120 aircraft in the air at a time. Food was the first requirement; coal and liquid fuel were other essentials.

Some idea of the immensity of the operation can be gained from the fact that between July 1948 and the following April, in more than 184,000 flights, some 1.5 million short tons of supplies were carried across 'the

Group Captain R. J. Cohen, senior air staff officer at No. 46 Group, RAF, during the Berlin Airlift, serving on exchange duty.

bridge'. Although the Russian blockade of the western sector of Berlin was lifted on 12 May 1949, the airlift was not officially discontinued until 30 September.

Members of the three RNZAF crews who flew Dakotas agreed it was the most difficult flying they had ever undertaken. On an average 12-hour working day they would make three flights from their base at Lubeck into Berlin. Their aircraft carried mainly coal, besides newsprint, cigarettes and a host of miscellaneous goods. On the return flights they often carried children suffering from malnutrition.

Flying conditions were frequently unpleasant and in the long winter months icing was a real problem. There was no room for error; there were always aircraft ahead and astern, above and below, too many, too close for anything but meticulously accurate flying.

At the end of a long day's flying in a congested air corridor, after fighting the weather and anxiously watching the Russian air activity, the New Zealanders were almost always incredibly weary. 'It was sheer drudgery', remarked one pilot afterwards, 'unrelieved by any of the lighter moments one knew in wartime. But it was a serious and important mission and no one complained.'

The RNZAF transport crews sent especially from the Dominion for this operation flew in the airlift for almost 13 months, though a six-month tour of duty had at first been envisaged. The three crews were: Flight Lieutenant C. J. Fraser, pilot, Flight Sergeant E. Saker, navigator, and Flight Sergeant N. Fitzgerald, wireless operator; Flight Lieutenant J. D. Phillips, pilot, Flight Lieutenant H. J. Hammond, DFC, navigator, and Flight Lieutenant S. H. Ward, wireless operator; Flight Lieutenant G. Haslop, pilot, Flight Lieutenant G. M. King, navigator, and Flying Officer R. Hicks, wireless operator.

Flight Sergeant E. G. Saker, awarded the Air Force Medal.

The contribution made by the New Zealand crews was recognised when Britain awarded Flight Sergeant Saker the Air Force Medal. Group Captain Cohen, who virtually controlled the operation of all British and American aircraft from the British zone into Berlin, was doubly honoured — by the British Government with the CBE and by the American Government with the US Legion of Merit. In 1986 Cohen, then a retired air commodore, still had in his possession the original draft of the operation order for the airlift which he had personally composed at short notice on the instruction of the Air Officer Commanding No. 46 Group, between 8 p.m. on one day and dawn the next morning. For 'Nugget' Cohen, foundation member of the RNZAF, one-time Deputy Chief of the Air Staff and Air Officer Commanding RNZAF Headquarters in London, the historic Berlin Airlift was his most challenging assignment. The responsibility he carried as Senior Air Staff Officer was considerable. Cohen's Air Force career had, in fact, centred around aircraft operations. It was a field in which he excelled and the airlift benefited from his talents.

More than ready for a respite, the RNZAF crews were repatriated soon after the blockade was lifted. But as the airlift did not end immediately a relief crew was sent from New Zealand at the RAF's request. It comprised Flying Officer D. O. Holloway (pilot), Flying Officer R. M. Dwerryhouse (navigator), and Flight Lieutenant R. W. Macadam (signaller). Other RNZAF transport crews en route to the United Kingdom to serve with the RAF's No. 24 (Commonwealth) Squadron on exchange duty also gained experience of the Berlin Airlift in its final phases.

The RNZAF's last sortie out of Berlin was made on 11 August 1949 by one of the exchange duty crews. On its completion the Commanding Officer of RAF Lubeck presented Flight Lieutenant C. L. Siegert with a handsome plaque to commemorate the event. Siegert saw to it that the trophy found a place of honour at Transport Wing Headquarters in Auckland. The plaque records that the RNZAF carried 1577 tons of coal into Berlin on 473 flights.

The Berlin Airlift was not an operation of war, although it cost the lives of some very gallant airmen, but some chroniclers have suggested that perhaps it averted another war in Europe. Be that as it may, it was a highly successful joint operation, with the air forces playing a dominant part. Mounted with the dramatic intensity of war, it was undeniably the most sustained large-scale air supply operation in the history of mankind.

The plaque presented to the RNZAF to commemorate its part in the Berlin Airlift.

A Dakota of No. 41 Squadron's A Flight flies low towards the dropping zone on a supply mission in support of ground forces in Malaya.

The Cold War Reaches Malaya

The Russian blockade which led to the mounting of the Berlin Airlift was but one manifestation of the tremendous pressure which Communist forces were exerting on the Western world at many points around the globe in 1948. A major conflict in China was pointing the way for a Communist victory there; authority was being challenged in Burma and Indonesia and there was bitter confrontation in Greece. That year the Cold War came also to Malaya.*

As the Malayan Communist Party mounted a campaign in a bid to add Malaya to the Soviet bloc, armed violence became an everyday occurrence on the Malay Peninsula. Led by Chin Peng, a capable but ruthless Chinese guerilla leader who had made a name for himself during the Japanese Occupation, terrorists attacked rubber plantations, tin mines and police outposts in a bid to undermine constituted authority.

On 16 June 1948 armed terrorists slaughtered three British rubber planters at Sungei Siput in Northern Malaya. Two days later the Malayan Government declared a state of emergency in Perak State and areas of Johore and on 24 June a state of emergency was declared by the Singapore Government. The Malay Peninsula, which had scarcely had time to recover from the effects of war and of the Japanese occupation, had now to endure 12 years of terrorism.

Responding to these developments, New Zealand began to consider its defence preparations with a growing sense of urgency. The Government was concerned to ensure that there was adequate defence in our immediate area, but accepted that in the event of global war between the Soviet Union and its allies and the United States and its allies, our major contribution would again be made in the Middle East. But the requirements of defence in our own area became more and more important and as the possibility of Communist thrusts down through the Malayan Archipelago towards the Australian perimeter was viewed with some alarm, talks were held between Great Britain, Australia and New Zealand which led to the ANZAM defence agreement (Australia, New Zealand and Malaya co-ordinated planning).

In 1949 the Communist victory in China was complete and insecurity in South-East Asia was worsening. The British Government began to press for assistance from Australia and New Zealand and in September 1949 an element of the RNZAF's No. 41 Squadron, known as A Flight, comprising three Dakotas, was dispatched to Singapore for service with Britain's Far East Air Force in support of security forces operating in Malaya. The flight was commanded by Flight Lieutenant R. A. Manners, and was based at RAF Station Changi, on Singapore Island. The air transport tasks saw the Dakotas carrying troops and supplies throughout the South-East Asian theatre, but an important role that was to be introduced to the RNZAF was the parachuting of supplies into the jungle for the support of the ground-based soldiers on patrol. Other missions included the escorting of flights of RAF Vampire jet fighters from Karachi across India, Burma and Thailand and survey flights over North Borneo.

By July 1951 A Flight of 41 Squadron had flown more than 488,000 miles on air routes extending from Ceylon to Hong Kong and Japan. In December of that year, after more than two years of operations from Singapore, the Dakotas were withdrawn to New Zealand where No. 41 Squadron was about to re-equip with new transport aircraft.

* The new Federation of Malaya came into existence officially on 1 February 1948, and with it came a revival of Communist interest, which had developed after the Japanese Occupation.

CHAPTER SIXTEEN

Postwar Restructuring

The war's end presented a unique opportunity for New Zealand to build an air force of the finest quality. Large numbers of highly trained and experienced aircrew were returning from overseas. Many, having grown up in the service of their country, had no roots in adult civilian life and, liking the life to which they had become accustomed, were keen to make the Air Force their future career. First-class tradesmen also offered themselves in vast numbers. But until the future role and likely commitments of the RNZAF could be defined, the size and shape of the postwar Service could only be guessed at and in reply to questions about career prospects, terms, rates of pay and conditions of service the Air Force had little to offer. All it could do was enlist men on interim engagements of two years.

Many, not surprisingly, would not accept service under these terms and so went into other avenues of employment, but some accepted the offer, so helping the RNZAF through a difficult and trying period. The 'interim' Air Force was to last for about two years.

It was inevitable that many New Zealanders who sought a career in the peacetime RNZAF turned instead to Britain and entered the Royal Air Force. By 1948, when the RNZAF was entering a more stable period, there were some 300 New Zealanders holding permanent or extended service commissions in the RAF.

Air Vice-Marshal Sir Arthur de T. Nevill. *S. P. Andrew*

Fortunately, there was good leadership where it was most needed. Air Vice-Marshal A. de T. Nevill, who succeeded Air Vice-Marshal Sir Leonard Isitt as Chief of Air Staff in 1946, was a sound and wise administrator, fearless, forthright and far sighted, and for these qualities he was widely respected. He had been a regular Service officer since graduating from the Royal Military College, Duntroon, in Australia, in 1919 and after service on the north-west frontier of India as an artillery officer, he had joined the New Zealand Permanent Air Force in 1930. He knew his own service intimately and had held most of the major command appointments before becoming Chief of the Air Staff. He was a man who invariably knew what he wanted and he was prepared to knock on any doors to ask for it. Nevertheless, he had no easy task in the period of uncertainty and confusion which resulted from the country's sudden transition from war to peace.

The paramount fear was that the Government, preoccupied with other issues, would allow the young Service to sink back into the state of inertia, starved of funds and equipment, which had overshadowed its formative years. Demobilisation was rapid. By the end of March 1946, 26,929 personnel had been discharged since VJ Day, out of a total 'home' strength of about 34,000.* Most of the stations established to meet the needs of war had been closed down. Stations remaining at the end of 1946 were: Lauthala Bay, Fiji, where No. 5 (Flying-Boat) Squadron was based, Whenuapai (Nos 40 and 41 Squadrons), Hobsonville (all technical training, a detached flight of No. 5 Squadron, and the Sunderland Transport Flight), Ohakea (operational flying) and Wigram (flying training). There were stores depots at Te Rapa, Te Awamutu, Mangaroa and Weedons, and a Headquarters Unit at Shelly Bay, Wellington. No. 14 (Fighter) Squadron was in Japan and No. 2 (BR) Squadron stationed at Ohakea, had been renamed No. 75 Squadron. No. 42 Squadron had been formally disbanded on 31 August 1946 and communications flying had become the responsibility of the General Purpose Flight at Ohakea.

The Planning Executive

Drafting a plan for the postwar reorganisation of the RNZAF was a task for a small, select team of officers, known as the Planning Executive, who were relieved of their existing duties early in 1946 and attached to the office of the Chief of Air Staff. The committee's chairman was 32-year-old Wing Commander I. G. Morrison and the other members were Squadron Leaders W. G. Woodward, N. A. Vear and W. J. R. Scollay. Morrison and Scollay were both officers of the flying branch and had a flair for planning and administration; Woodward had joined the Air Force through the university scheme,

* The figure does not include RNZAF personnel in the RAF and Canada.

trained as a pilot, and was then moving up in the Engineering Branch; Vear was a senior engineer officer. Appointed while Isitt was still CAS, they were to serve his successor, Air Vice-Marshal Nevill, very well indeed as his 'think tank'. The Planning Executive was directly responsible to the Chief of Air Staff who, after Air Board agreement, made appropriate recommendations to Cabinet on such matters as officer and airman establishments, rank structure, pay and conditions of service. The Planning Executive worked at its task for almost three years. A uniform three-Service pay structure was one direct result of its deliberations; another was the emergence of the Secretarial Branch. Scollay sat in on the discussions of the RAAF's post-hostilities committee in Melbourne and the postwar planning of the RAF was examined closely, parts which were applicable being adapted and used to suit New Zealand conditions. The Executive also had to decide on the disposal of a large number of wartime aircraft and on what types should be retained in the peacetime Service.

With the foundations laid, the reshaping of the Service went ahead. The Regular Air Force was reorganised on a peacetime basis, the Territorial Air Force was reconstituted and an Air Force Reserve formed. The Women's Auxiliary Air Force, which had been formed in 1941 to release men for more active duty, was retained. The Air Training Corps, also formed in 1941 to train youths for eventual service in the RNZAF, was also retained, divided into two parts: the town squadrons and the school units (which formed part of the Cadet Corps of New Zealand secondary schools).

The first two years after the war ended were an interesting period for the RNZAF air transport service (which included the Sunderland Flight), for besides meeting the

No. 5 Squadron Sunderlands at RNZAF maritime base, Lauthala Bay.

The Air Training Corps was retained in the postwar RNZAF with town squadrons and school units. Here cadets familiarise themselves with a Service hand-held aerial camera.

normal requirements of the Air Force it also provided air transport schedules for the civilian population, operating what amounted to an airline service within New Zealand and the Pacific Islands until the emerging Government-owned New Zealand National Airways Corporation could take over.

The quasi-civil internal services began officially on 1 October 1945 and the internal Dakota schedules were run largely by No. 40 Squadron until NAC took them over on 2 June 1947. (For the year ending 31 March 1947, 90 per cent of the passengers carried by 40 Squadron and the Sunderland Flight were civilians.) No. 42 Squadron operated an internal mail service between Wellington and Auckland, using Dominie aircraft (it had begun in April 1944 with a Harvard but the volume soon called for larger aircraft), and this schedule allowed for stops at New Plymouth and Rukuhia if required. Service passengers were carried if space was available. From 10 February 1947 the RNZAF also operated a freight service between Paraparaumu and Woodbourne under contract to the New Zealand Railways. Both these services were initially absorbed by NAC.*

Squadron passed and this unit was disbanded on 31 October 1947. No. 41 Squadron maintained a regular courier service to Japan, where 14 Squadron was serving, from March 1946 to April 1948. During that period 104 flights were made, the aircraft carrying 1477 passengers, 114,420 lb of freight and 248,711 lb of mail over 1,323,872 miles.

The Forest Fire Patrol and Aerial Topdressing

Two other national projects of prime importance to New Zealand occupied the attention of the RNZAF in the immediate postwar period. One was the aerial patrolling of the vast, man-made pine forests of the Taupo-Kaingaroa-Rotorua area to prevent fire from destroying one of the country's richest assets. The other was a series of experiments in aerial topdressing of marginal lands which paved the way for the birth of a new industry.

The Forest Fire Patrol established at Rotorua's old grass airfield in 1946 was a temporary expedient offered by the Air Force to provide an 'air arm' for the New Zealand State Forest Service until a civilian counterpart could be found. Established with two aircraft, two pilots and a

An Airspeed Oxford of No. 42 Squadron on the ground at Ohakea in 1951.

An Auster of No. 42 Squadron in the early 1950s.

NAC also acquired the Air Force's Dakotas, Lodestars, Dominies and Sunderlands which were given a new and distinctive livery. Aircrew members of the RNZAF transport units and Air Force technicians had been told as far back as late 1945 that they could apply for service in the proposed national airline and many had done so.

As NAC gradually took over responsibility for civilian air transport requirements, the Service's need of 40

* A new, freight-carrying airline, Straits Air Freight Express Ltd, later to become Safe Air Ltd, was formed in 1951 to operate the rail-air service under contract to the New Zealand Railways.

small servicing party, its field of operations was the centre of the North Island of New Zealand, comprising 750,000 acres of exotic and indigenous forest. The nation then drew most of its timber from these forests, the biggest single block being the Kaingaroa State Forest of 350,000 acres.

Until the introduction of the aerial patrol, disaster had come to the forests from time to time in the form of devastating fires which had swept uncontrolled through vast areas of plantation. The Taupo fire of 1946 which destroyed 30,000 acres of forest strengthened the hand of the State Forest Service and enabled it to demand new measures, manpower and equipment. Among these measures was the Forest and Rural Fires Act 1947, which enabled the Forest Service to create a large-scale firefighting organisation designed to prevent fires rather than combat them. Firebreaks were increased and widened, new forest lookout stations were erected and manned, and the aerial patrol was placed on a better footing.

Long before this — in the summer of 1944–45 — a patrol had been offered by RNZAF Station Rukuhia in the Waikato district, an arrangement having been made with the Forest Service that an Airspeed Oxford and a pilot would be detailed for a flight if required. But this was far from satisfactory and it was subsequently agreed that the air arm of a forest firefighting organisation could be efficient only if based at Rotorua, in the heart of the afforestation area, located there permanently during the dry season and equipped with suitable light aircraft.

Consequently it was decided to establish an RNZAF unit at Rotorua and arrangements were made for the de Havilland Aircraft Company to modify aircraft for this task. As an interim measure, Flight Lieutenant J. R. Claydon was dispatched to Rotorua in October 1946 to provide a service with one Harvard aircraft. The experiment proved the value of aircraft in this role and a properly constituted RNZAF detachment from Ohakea moved into Rotorua with two Tiger Moths and one Fox Moth, aircraft later replaced by Auster J5s. The RNZAF continued to provide an Aerial Fire Patrol each summer until early 1952 when a civilian operator took over the task.

Aerial topdressing of New Zealand's marginal lands was first postulated by a New Zealand farmer in 1926 but the idea met with scepticism. The first aerial seeding was carried out by the Public Works Department at Waipapakauri in March 1941 when lupin seeds were sown for sand dune reclamation. Further trials were conducted by the PWD (later renamed the Ministry of Works and Development) in 1946 and again in 1947, not only of seeding but also of fertiliser dropping on the airfield at Rongotai. At this stage the Department of Agriculture (renamed the Ministry of Agriculture and Fisheries) and the Soil Conservation and Rivers Control Council began to show interest in the project. A meeting of interested parties in

A Tiger Moth of the RNZAF Aerial Fire Patrol in 1947.

November 1947 decided to carry out two aerial seed sowing trials and one aerial topdressing trial. In April the following year, when the RNZAF was asked to carry out the aerial topdressing trial, the Air Board agreed and allocated a Grumman Avenger aircraft.

The RNZAF's part in the trials was directed by Group Captain W. C. Sheen, an RAF officer on loan, then Deputy Chief of the Air Staff. Group Captain Sheen played a major role in the developmental work and it was largely due to his enthusiasm that the experiments were so successful.

Ohakea was selected for the initial experiment. The Avenger aircraft used was fitted with an auxiliary petrol tank converted into a hopper which was installed in the bomb bay. It was loaded to carry 2240 lb of fertiliser. (The load was reduced to 1500 lb for later trials.) On 5 September 1948 the RNZAF dropped its first test load and the following day official trials began, in which loads of superphosphate were dropped alongside and across the runways at Ohakea. The trials were highly successful and more were carried out on 21 September.

Following the success of the Ohakea experiments

Aerial topdressing trials with an Avenger especially converted for the role.

arrangements were made for the first full-scale field trials. A suitable farm property was selected at Te Mata, near Raglan, and on 12 October 1948 Operation Topdress commenced. Some 2⅓ tons of fertiliser were dropped over an area of 27 acres. In this trial, as in all subsequent trials in the field, the RNZAF established a system of point-to-point, air-to-ground and ground-to-air radio communications which, after a few initial problems had been eradicated, functioned satisfactorily. The next trial, in March 1949, involved a 50 acre property at Sanson, near Ohakea. For this experiment the hopper was modified to dispense clover seed as well as fertiliser.

To further facilitate experiments in aerial topdressing No. 41 (Transport) Squadron was reorganised in April 1949 to include a Research and Development Flight. Commanded by Squadron Leader S. G. Quill, DFC, this flight was to be responsible for future aerial topdressing experiments. With the experience gained in the earlier small trials behind it, the Air Force now proceeded to carry out the first full-scale topdressing operation, to evaluate a miniature aerial topdressing service. A series of sections in the Wairarapa district with a total area of 1000 acres was selected. The aircraft complement of the Research and Development Flight was increased to three Avengers and attendant services were likewise increased.

Hood Aerodrome, Masterton, was chosen as the base site. The first drops were made on 3 May 1949 and operations continued until 21 May. The total flying time completed during Operation Topdress III was 48 hours 20 minutes, in which time a total of 125 tons of superphosphate and 12 tons of lime was dropped. In addition, grass and clover seed was mixed with the fertilisers on some of the drops.

One further trial was conducted by the Air Force, this time in the Ohariu Valley near Wellington for the benefit of watching Members of Parliament. Two Avengers topdressed an area of 50 acres successfully despite difficult wind conditions. After the Ohariu trial, the RNZAF did not make any further field trials but some tests were made with Miles Aerovans at Whenuapai in 1950. The Air Force withdrew as commercial operators stepped in.

A de Havilland Mosquito fighter bomber, Mk VI.

Mosquitoes on the airfield at Kaitaia during Exercise Hardtack. *Evening Post*

CHAPTER SEVENTEEN

A New No. 75 Squadron

In 1946 it was still the belief in Government circles that New Zealand's future lay with Britain and the Commonwealth, and this meant common aims, common strategy and standardised equipment. There was still uncertainty about the role of the peacetime Air Force and its likely commitments, but as an interim measure, 80 de Havilland Mosquito fighter-bombers were ordered from Britain. It was envisaged at the time that the Mosquitoes would arm three squadrons, one of which would be deployed to Fiji, but because of manning difficulties only one squadron was formed. The Mosquitoes were purchased for £3,000 each; taking into account extra engines and a quantity of spares, the total purchase cost £700,000.

On 11 October 1946 it was announced that the Air Ministry had consented to the designation 'No. 75 Squadron' being passed to the RNZAF to commemorate the remarkable contribution of No. 75 (New Zealand) Bomber Squadron in the Royal Air Force in the Second World War. Accordingly, said the Department, No. 2 (Bomber Reconnaissance) Squadron at Ohakea would thereafter be known as No. 75 Squadron.

Wing Commander J. E. Watts, AFC, was appointed to command the new 75 Squadron. His first task was to arrange conversion training on Mosquitoes for pilots and navigators who would take part in ferrying the aircraft to New Zealand.

The squadron's first four Mosquitoes, delivered by RNZAF crews in November, were surplus Australian dual-control aircraft bought for this purpose. The first four of the United Kingdom consignment were flown out by RAF crews in January 1947 and that same month the first 75 Squadron ferry crews departed for Britain. On 28 March the first Mosquito ferried by a New Zealand crew arrived from Britain, captained by Squadron Leader C. H. Baigent, DSO, DFC and Bar.* Two other Mosquitoes flown by RAF crews had accompanied him on the 17-day journey. During the year 54 Mark VI Mosquitoes arrived from Great Britain and a further four dual-control Mosquitoes from Australia were ferried across the Tasman.

The Mosquito ferry was not accomplished without some casualties. On 29 March 1947 one Mosquito and its RAF crew were lost in the Gulf of Carpentaria during a tropical storm, and on 19 August 1947 another Mosquito flown by an RAF pilot and an RNZAF navigator crashed at Darwin when an insect became lodged in the pitot head during take-off. The aircraft overran the end of the runway and was damaged beyond repair but the crew escaped injury. On 27 August 1947 Flight Lieutenant R. B. Burney, AFC, and Flight Lieutenant B. J. Beadle were reported missing between Mauripur and Santa Cruz in bad weather. A search was made but they were not seen again. On 26 November, Flight Lieutenant F. Adams and Flying Officer M. J. Fry were involved in an accident at Mingaladon, Rangoon. After take-off the pilot detected an undercarriage fault and decided to land. The Mosquito swung into a monsoon ditch, caught fire and was destroyed, but both pilot and navigator escaped with burns. Apart from the losses there were a few incidents and several delays while crews waited for spares to arrive. Round trips, which had been calculated to take from six to eight weeks, sometimes took two and a half to three months.

By late 1947 No. 75 Squadron was settling down and a training programme was being instituted. The squadron was also exercising with HMNZS *Bellona* in a variety of aircraft-versus-ship attacks. Many air displays were being staged throughout New Zealand at this time and the Gloster Meteor, flown mostly by Squadron Leader R. M. McKay, was a great attraction in aerobatics and high-speed runs, while the Mosquito crews, by now quite at home in their aircraft, were able to stage some thrilling displays of aerobatics and formation flying for the public.

In December 1948 the last Mosquito from England flew into Ohakea and as the year 1949 dawned, No. 75 Squadron's status had changed from that of a mobile ferrying force to a well-established and reasonably static unit. Seventy-six of the 80 Mosquitoes ordered had been delivered, a wide programme of training was in progress and 1163 hours of flying had been logged during the year. Squadron Leader Baigent, who had taken command of the squadron in April 1947, had moulded the men into the sort of operational team he had known in the Northern Hemisphere when he commanded the original 75 (New Zealand) Bomber Squadron.

During 1949 the squadron extended its operations to include the Chatham Islands and the Kermadec Group and in May 1950 a detachment flew to Nandi in Fiji and later to Nausori where crews were met by Wing Commander R. Webb, DSO, DFC, the Officer Commanding the RNZAF's Lauthala Bay station. A second deployment was made to Fiji on 25 August by three aircraft from the squadron. The deployment ended on 8 September.

* In accordance with normal practice at the time he had dropped rank since his wartime service.

Another exercise in 1950 saw squadron Mosquitoes, offered the barque *Lutterworth* as a target, sink the vessel off Baring Head with rocket fire. In January 1952 the hulk of the former ferry *Arahura* was sunk in Cook Strait by eight Mosquitoes attacking a supposed enemy merchantman attempting to make passage through the strait; this exercise was dubbed Operation Scuttle Two. The squadron deployed to Kaitaia, then a little-used aerodrome, in March that year for Exercise Hardtack and operated for two weeks from a tented headquarters and camp on the edge of the airfield.

In April 1952 the squadron was formally notified that training on Mosquitoes was to cease, that jet-propelled de Havilland Vampire fighter aircraft were to replace the Mosquitoes and that while 75 Squadron would remain on the RNZAF's Order of Battle the Mosquitoes would go into storage at Woodbourne and the personnel would be attached to No. 14 Squadron. Squadron Leader E. C. Gartrell OBE, DFC, who had commanded No. 75 Squadron since May 1950, flew the last Mosquito, NZ2306, to Woodbourne on 24 April.

Squadron Leader C. H. Baigent, DSO, DFC, appointed commanding officer, No. 75 Squadron, Ohakea, April 1947.

Squadron Leader E. C. Gartrell, OBE, DFC, who commanded No. 75 Squadron in the final phase of its Mosquito era, 1950–52.

A Mosquito of No. 75 Squadron in flight.

A Vampire Mk FB.52 jet fighter. The Vampire was the first jet aircraft to go into squadron service with the RNZAF.

CHAPTER EIGHTEEN

No. 14 Squadron in Cyprus

After a meeting of Commonwealth Ministers in June 1951, New Zealand decided to contribute to the strength of the Middle East Defence Forces and No. 14 (Fighter) Squadron, after training on the new de Havilland Vampires at Ohakea, moved to Cyprus in October 1952 to serve there in a garrison role. Leaving their Vampires behind as replacement aircraft for No. 75 Squadron, they flew to the Middle East in the RNZAF's new Handley Page Hastings long-range transports.* An advance party, under Flight Lieutenant H. A. Farrar, the squadron engineering officer, had left some weeks earlier to establish the squadron in their new home at Nicosia, the main civil/military airfield of Cyprus.

The main party, under the command of Squadron Leader S. M. Hope, arrived from 7 October onwards, and squadron pilots were airborne in their new aircraft three days later. Hope was an Army officer who had transferred to the Air Force in 1942, learned to fly, and flown fighters in the Pacific and with 14 Squadron in Japan. Known as a painstaking and meticulous junior staff officer who excelled on the parade ground, he was chosen as A Flight Commander of 14 Squadron when the unit was first reformed with Vampires, but shortly thereafter took over as Squadron Commander. The Cyprus deployment gave him the opportunity and challenge to develop as a leader, and his career went forward naturally from that point. To the MBE bestowed on him in the 1950 New Year Honours was added the AFC in 1953 in recognition of his leadership in the Middle East.

Hope was supported by Flight Lieutenants T. R. Rabone and J. D. Waugh, both of whom had also seen fighter service in the Pacific and Japan. Waugh had completed a jet conversion course and gunnery instructor training in the United Kingdom, before the Vampires were delivered to New Zealand. The remainder of the pilots were postwar trained, and without any overseas experience.

Flying Mark 9 Vampires hired from the RAF, the squadron became part of No. 205 Group, RAF, under the operational command of HQ Middle East Air Force located at Abu Sueir in the Suez Canal Zone. Operating out of RAF bases in Egypt, Iraq, Jordan, Malta and Cyprus, as well as staging posts throughout the Middle East, the squadrons of No. 205 Group flew Vampires and various marks of Meteor in the day fighter/ground attack, photo reconnaissance and night fighter roles, and 14 Squadron very quickly found itself competing against the much more experienced RAF and RAAF squadrons. (No. 78 Wing, RAAF, was already based at Malta with two Vampire squadrons, many of whose pilots had recently fought in Korea.)

While 14 Squadron became the only resident squadron at Nicosia, the base was host to a constant stream of day fighter/ground attack squadrons attending No. 26 Armament Practice School, also located at Nicosia. As a consequence the squadron was able to gain much-needed experience of operating in wing-sized formations, and of measuring their skills against other units. The majority of the squadron were very young, and they were quick to learn. In short order, they were able to rise to the level of the best units in the group, and in at least one field, rocket firing, they had no equal.

Mobility was a strong factor in operations in the Middle East. 14 Squadron aircraft routinely deployed to the Canal zone (a 50-minute overwater flight) where they flew practice air-to-air interceptions and operated with the RAF squadrons there, and from time to time the squadron deployed further afield on specific exercises. Two of the more demanding and exciting of these were Long Trek Two in May-June 1953 and, in August of that year, DXM 53, as the defence exercise of Malta was nicknamed.

Squadron Leader S. M. Hope, the first officer commanding No. 14 Squadron in Cyprus.

* See Chapter 19, 'A Decade of Change'.

The Prime Minister of New Zealand, the Rt Hon. S. G. Holland (fifth from left), with No. 14 Squadron members at Nicosia, Cyprus.

In Long Trek Two the squadron deployed from Cyprus to East Africa in three independent flights of five aircraft each in order to mount flying displays as part of Queen Elizabeth's Coronation celebrations. Most Middle East squadrons had a formation aerobatic display team, but 14 Squadron had not one but two full teams at that stage, and so deployed one formation aerobatic team each to Uganda and Kenya and a display flight to Tanganyika, now part of Tanzania.

The deployments involved very long and demanding legs for single-engined, single-seat aircraft without navigation aids or auto-pilot assistance. The route roughly followed the old Imperial Airways flying-boat route up the Nile to its source in Uganda, and then turned east to Kenya and the furthest destination, Dar-es-Salaam, some 3330 miles from Cyprus. The longest leg (over two hours) was between Khartoum and Juba, a strip hacked out of the rainforest at the southern extremity of the navigable Nile to where the aircraft fuel had been lightered in 4-gallon tins. That particular leg usually meant flying through the inter-tropic convergent zone, and with the aircraft flying in pairs for mutual navigation support and safety, the pilots found themselves having to fly close formation in cloud, and with iced-up instruments, for up to 45 minutes. Despite the difficulties involved, the deployment was entirely successful and all aircraft and supporting ground crews returned to Cyprus some three weeks after setting out.

DXM 53, while not so demanding in deployment terms, was one of the most exciting Middle East exercises in which the squadron participated. Deploying to Malta via El Adem in Libya, the squadron operated alongside the RAAF wing in the day air defence of the island against carrier-borne attacks by Banshees, Panthers and Skyraiders of the US Sixth Fleet. Many of the attacks were below radar level, with the only warning being relayed from lookouts around the coast. Very high states of alert and rapid scrambles were demanded of the defenders and much of the air-to-air fighting took place at low level. Quick turnarounds placed heavy demands on the 30 ground crew, but despite this they kept at least 12 out of their 13 aircraft at Malta on readiness, and the squadron flew 125 sorties in 96 hours during the exercise.

In May 1954 Squadron Leader Hope was succeeded by Squadron Leader N. H. Bright who had been commanding 75 Squadron in New Zealand. His stay in the Middle East was to be relatively short, however, for the year 1954 marked a turning point in the strategic thinking of Britain, Australia and New Zealand. Whereas in 1951 the Middle East was still seen as an area of primary strategic importance to the British Commonwealth, by 1954 the focus of interest had shifted to the Far East where Communist-inspired insurgency was calling for stepped-up counter-measures.

In late 1954 Britain asked both Australia and New Zealand to contribute forces to a Commonwealth Strategic Reserve to be maintained in the Far East. The result was a decision to transfer No. 14 Squadron from Cyprus to Singapore. The New Zealand Prime Minister, Mr Sidney Holland, gave notice of this intention publicly in February 1955 after a visit to Malaya and he formally announced the move on 24 March. The following month, still under the command of Squadron Leader Bright, No. 14 Squadron arrived at RAF Station Tengah, on Singapore Island, to serve alongside squadrons of the Far East Air Force.

CHAPTER NINETEEN

A Decade of Change

Serious shortfalls in certain ground trades — a consequence of the rapid demobilisation which had followed the end of the war — were causing manning difficulties in 1949 and 1950 and to remedy the situation the RNZAF turned to the United Kingdom. Former members of the RAF were interviewed by staff officers at RNZAF Headquarters in London and experienced tradesmen interested in emigrating to New Zealand were recruited.

During the latter half of 1950 the RNZAF also recruited in the United Kingdom 100 airwomen, all of whom had served for a minimum of two years in one of the British Armed Forces. They sailed from Southampton in the *Atlantis* in April 1951. Because one of the draft became ill and left the ship at Malta, the contingent was 99-strong when it arrived in New Zealand and through the years this group of airwomen has been known as 'the Ninety-Nine'. They went to RNZAF Station Woodbourne for a month's indoctrination and general service training and were then absorbed into the Service where they generally fitted in remarkably well and were valued for their efficiency.

A prominent member of the WAAF in 1951 was Warrant Officer C. F. Bryers, regarded by her senior officers as an outstanding airwoman. Frances Bryers had joined the WAAF in 1942. An excellent sportswoman, this Maori warrant officer enjoyed great popularity with all ranks. She was one of the contingent especially chosen to represent the RNZAF at the Victory Parade in London in 1946. Warrant Officer Bryers was made an MBE in 1961 and an accommodation wing at Wigram was named in her honour in 1984.

In 1954 the Queen gave approval to the title of the WAAF being changed to Women's Royal New Zealand Air Force. By then it had become a vital part of the Service with members serving at all RNZAF stations, including Lauthala Bay in Fiji.

Warrant Officer C. F. Bryers, MBE.

Some of the UK-recruited WAAFs, pictured on board the *Atlantis* soon after their arrival in Wellington in 1951.

The Mustang fighter which saw service with the four squadrons of the Territorial Air Force.

A Handley Page Hastings transport.

A Bristol Freighter.

A de Havilland Devon.

A Vampire engine change.

A Sunderland at Lauthala Bay.

Aircraft Re-equipment Programme

In its first postwar, major re-equipment programme, the RNZAF in 1951-52 began taking delivery of new aircraft, all of which were to give long and notable service. Total deliveries, which for some types extended over six years, included four Handley Page Hastings four-engined, long-range transports, 12 Bristol 170 Freighter Mark 31 twin-engined medium-range transports, 46 de Havilland Vampire jet fighter/ground attack aircraft, 11 Vampire jet trainers and 28 twin-engined de Havilland Devons* for use as navigation and radio operator trainers at Wigram, and as light transports. In addition, 16 reconditioned Short Sunderland flying-boats went into service in the maritime reconnaissance role. It was during this period that the four fighter squadrons of the Territorial Air Force (TAF) were re-equipped with Mustangs which, purchased new, had been cocooned and stored since the end of the war.

The aircraft re-equipment programme had been initiated during Air Vice-Marshal Nevill's four and a half years' term of office — a longer term than any Chief of Air Staff before him or since. Air Vice-Marshal Nevill, knighted for his able leadership in difficult times, was appointed Director of Civil Aviation in 1956.

It fell to his successor, Air Vice-Marshal D. V. Carnegie, CB, CBE, AFC, to implement the re-equipment programme. Carnegie took office in January 1951, on loan from the RAF, following the prewar pattern which had begun with Cochrane. When Nevill retired that month, after steering the RNZAF through the interim period into more stable times, the aircraft purchases had been approved by Government, flying training had been resumed at Wigram, an Officers School had been formed

* Two early Devons, NZ 1801 and NZ 1802, had been delivered in 1948.

A Mustang of No. 4 (Otago) Squadron, Territorial Air Force, over Otago Harbour.

A No. 42 Squadron Dakota towing drogue for the Royal New Zealand Navy gunnery practice.

Crowd thrillers three! Harvards of the Central Flying School's aerobatic team, the Red Checkers, 1969.

Opposite page: Territorial Air Force Mustangs on the line at Ohakea during an annual camp. The distinctive chequer-board markings were applied in the provincial rugby colours of each squadron's home province.

at Whenuapai to give administrative training to young officers and RNZAF Station Woodbourne had been reopened to accommodate the Boy Entrant School, transferred from Wigram, and No. 1 Repair Depot, from Hobsonville.

The Air Force which emerged in the early 1950s under Air Vice-Marshal Carnegie had a strength, excluding territorial units, of about 4000 men and women. It was compact, well organised and capable of rapid expansion in time of need. Inevitably it continued to follow the established pattern of the Royal Air Force in training, organisation, discipline and customs.

By the second half of 1953 the RNZAF had rid itself of much of its wartime equipment — with the notable exceptions of a few Dakotas and the pilot-training fleet of Harvards — and with many of its new aircraft in service was in comparatively good heart. At Lauthala Bay, Fiji, No. 5 Squadron was engaged in maritime reconnaissance and anti-submarine warfare training with the first of its Sunderlands, which had been delivered in June that year. No. 41 Squadron at Whenuapai was operating Hastings aircraft in the long-range transport role and Bristol Freighters to meet medium-range requirements. No. 14 Squadron was serving in the Middle East and No. 75 Squadron was training on Vampires at Ohakea. No. 42 Squadron, reformed on 15 March 1950, with a variety of aircraft types, was operating from Ohakea in the communications role and was progressively introducing both Dakotas and new Devon aircraft into service.

Compulsory military training (CMT) had been introduced in 1949 following a national referendum that year which gave the Government a positive mandate for its action, and at RNZAF Station Taieri the Initial Training School was providing initial instruction for aircrew, both Regular and CMT intakes. At Wigram the Flying Training School had increased its output of pilots, Harvard trained, for the RNZAF, RAF and TAF. The Air Navigation and Signals School was busy training with new Devons.

These Regular units were supported by five squadrons of the TAF. Four territorial fighter squadrons had been formed in late 1948, following a Government decision to establish reserves of air and ground personnel to support the Regular forces. Nos 1, 2, 3 and 4 Squadrons were established in Auckland, Wellington, Christchurch and Dunedin respectively in the fighter/ground-attack role and flying began late in December. Initially Tiger Moths and then Harvards were used, but in 1951 and 1952 the squadrons were issued with the North American P-51 Mustang fighters which had been delivered to the RNZAF in 1945 but had not gone into service because of the sudden end to hostilities that year. A fifth TAF unit, No. 6 (Maritime) Squadron, had been activated at Hobsonville in April 1952, initially with Catalinas and Auster floatplanes. In 1954 it began operating Sunderlands.

The TAF fighter pilots were delighted with the fast, powerful Merlin-engined Mustang, without doubt one of the finest and most versatile fighter aircraft ever built, but weekend local flying and an annual training camp at an RNZAF station did not allow the territorial pilots to amass much experience on the type and there were several fatal accidents, enough to cause concern within the Air Staff.

The London to Christchurch Air Race

In October 1953 Hastings NZ5804 of No. 41 Squadron was the RNZAF's entrant in the transport section of the London to Christchurch Air Race. Wing Commander R. F. Watson was in command and he had several experienced crew members with him. About an hour out from Negombo, Ceylon (now Sri Lanka), when the aircraft was off the coast of India and darkness had fallen, the Hastings encountered a bad monsoon storm. Squadron Leader C. L. Siegert was flying at this stage and he worked hard to keep the aircraft under control in severe turbulence. It was than that the port inner engine failed, but Siegert

Wing Commander R. F. Watson.

Squadron Leader C. L. Siegert.

feathered the propeller on the faulty engine and carried on through the storm. Calling the control tower at Negombo, he learned that the storm covered the airfield and visibility was practically nil. Despite the conditions, however, the aircraft would have to land there on three engines as the fuel supply did not permit diversion to another field. The runway was covered in water to a depth of 2 inches, and when the Hastings touched down — after two overshoots — the impact of water streaming back from the undercarriage buckled the flaps.

Checks made after the landing revealed fragments of metal in the oil, indicating a sleeve-valve failure. Sealed engines had been a condition of the contest; by the rules they were out of the race. The crews could only wait for another engine to be flown out from New Zealand and accept the disappointment of flying back to a country where the London to Christchurch Air Race was already history.

Presentation of the Queen's Colour

During the 1953–54 Royal Tour of New Zealand — the first visit to this country by Queen Elizabeth II and her husband, the Duke of Edinburgh — the Queen personally presented her Colour* to the RNZAF. The presentation was made at Whenuapai during a ceremonial parade on 28 December 1953.

Six hundred officers, airmen and airwomen took part, and besides the invited guests thousands of onlookers observed the scene from tiered grandstands erected for the occasion. Cadets of the Air Training Corps lined the driveway. On the parade ground the representative squadrons, the Band of the RNZAF and the Colour Escort Squadron stood at attention. After members of the Air Board and their wives and the Station Commander, Wing Commander T. J. MacLean de Lange, DFC, and Mrs de Lange had been presented to the Royal couple, the Queen,

* The Queen's Colour is described officially as 'a blue field, surrounded by a ring of New Zealand fern leaf in gold, with the Union Jack in the top left corner, the four main stars of the Southern Cross in the top right corner and the Royal Cipher surmounted by a crown in the centre.'

A Hastings of No. 40 Squadron at Nausori (Fiji) in March 1965. The Hastings gave the RNZAF its first true long-range transport capability.

Queen Elizabeth II presents her Colour to the RNZAF. The Colour Bearer is Flying Officer M. F. McD. Palmer. With the Queen is Air Vice-Marshal D. V. Carnegie, Chief of the Air Staff.

accompanied by the Duke, inspected the parade. The Colour was consecrated by the senior chaplain, the Reverend A. R. Allerton, MC, after which the Chief of Air Staff, Air Vice-Marshal Carnegie, handed it to the Queen. The Colour Bearer, Flying Officer M. F. McD. Palmer, received the Colour from Her Majesty who concluded her address with these words: 'I give you my Colour in the knowledge that you will guard it well; in the belief that it will always remind you of the trust that I place in my Royal New Zealand Air Force. During your future service you will, I know, be always worthy of that trust and will always strive to maintain the high and honourable standards of which this Colour is a symbol.'

For the 1953–54 Royal Tour the RNZAF created a 'Queen's Flight' of transport aircraft within No. 42 Squadron. A highly polished VIP Dakota, NZ3551, specially refurbished for the occasion, was captained by Squadron Leader Siegert, who was decorated with the MVO by the Queen personally for services to the Monarch.* The plaques which adorned the interior of this aircraft, all replicas of squadron badges, were Air Vice-Marshal Carnegie's personal contribution to the Royal aircraft. (A second Dakota was prepared as a back up and these two aircraft were to serve in the VIP transport role with 42 Squadron until 1977.)

* Air Vice-Marshal C. L. Siegert, CB, CBE, MVO, DFC, AFC, Chief of Air Staff, RNZAF, 1976–79.

No. 40 Squadron Reactivated

The RNZAF's air transport organisation underwent a significant change in December 1954. No. 41 Squadron, which was operating the Hastings in the long-range field and the Bristol Freighters in the medium-distance role, was reorganised into two separate squadrons. No. 40 Squadron was reactivated and allocated the four Hastings aircraft. Flight Lieutenant R. B. Bolt, DFC,+ who had flown on heavy bomber and Pathfinder Force missions during the Second World War, was promoted to squadron leader and given command of the new unit. The reactivation of No. 40 Squadron left 41 Squadron with the Bristol Freighter component.

In New Zealand and the Pacific region search and rescue missions and air ambulance flights were flown by Hastings, Bristol Freighters and Sunderlands. These aircraft were also called on for ocean searches for missing vessels and for relief missions after natural disasters, such

+ Air Marshal Sir Richard Bolt, KBE, CB, DFC, AFC, Chief of Air Staff, RNZAF, 1974–76, Chief of the Defence Staff 1976–81. He was the first member of the RNZAF to reach the rank of air marshal in that Service. Pioneer pilot and aero engineer Mr George Bolt was his father.

The Queen disembarks from Dakota NZ3551 at Kaikohe during the 1953–54 Royal Tour. Following the Queen are the Duke of Edinburgh and the New Zealand Prime Minister, the Rt Hon. S. G. Holland.

as hurricanes. Some of the flights undertaken called for airmanship of the highest order. One of the longest mercy flights ever undertaken by No. 5 Squadron was made in 1950 — a 4600 mile ocean crossing to Penrhyn Island, in the extreme north of the Cook Group, to attempt to save the life of a woman who was seriously ill after a complicated childbirth. The aircraft, which carried a European doctor and senior nurse and a Fijian medical practitioner, flew to Samoa on the first day, to Aitutaki on the second and the final 600 miles to Penrhyn on the third. The flying-boat landed on a lagoon 12 miles offshore and the medical team completed the journey by launch. In spite of very limited facilities, major surgery was undertaken and when the team left the next day the patient was already showing signs of recovery.

This incident was a highlight of the Catalina era, but No. 5 Squadron responded to many similar calls during the many years that it flew Sunderlands. Gradually the Island communities came to depend on the RNZAF in Fiji and, despite its primary role of maritime reconnaissance and anti-submarine warfare training, 5 Squadron became widely known and respected throughout the Pacific Islands for its humanitarian and relief operations.

The Hastings aircraft of No. 40 Squadron also ranged far afield. They regularly flew the 13,750 mile route between Whenuapai in New Zealand and Lyneham in Wiltshire, England, and they undertook the air dropping of supplies to isolated offshore islands such as Campbell Island, the Kermadecs and the Chatham Islands. Together with the requirements of the Armed Forces for troop carrying and the resupplying of bases in both hemispheres, this was a formidable workload for a squadron of only four aircraft and when Hastings NZ5804 was damaged and had to be written off in an accident at Darwin in 1955 three aircraft shared this burden for a further 10 years.

Territorial Units Disbanded

By the mid 1950s the old strategic concepts were changing and defence planners were agreed that there was little chance of New Zealand again having a breathing spell in which to prepare for war. The old concept of a Regular nucleus with reserves capable of rapid mobilisation in time of war was unrealistic and — owing to the ever-increasing cost of modern aircraft — uneconomic. The emphasis was placed on a well-equipped, highly trained Regular Air Force which could be rapidly deployed to any theatre.

In late 1955 Air Vice-Marshal W. H. Merton, CB, OBE, RAF, who had succeeded Air Vice-Marshal Carnegie as Chief of the Air Staff the previous year, directed that the Mustangs be withdrawn from territorial squadron service. The pilots trained instead on Harvards until 1 August 1957 when the four territorial fighter squadrons and No. 6 (Maritime) Squadron, were disbanded.

CMT came to an end for the RNZAF in 1957 and was abolished officially in 1958. The Air Force and the Navy had not derived much benefit from the scheme, principally because of the difficulty of producing skilled servicemen in the more important trade categories when the maximum training period in any one year was only 14 weeks. The disbandment of the TAF squadrons came as a disappointment to air and ground crew members alike and particularly to the squadron commanders who were proud of their units and, faithful to the memory of their Regular days, had given devoted service.

The year 1953 saw the retirement of Mr T. A. Barrow, CBE, JP, who, as Air Secretary, had been Permanent Head of the Air Department and a member of the Air Board since 1937. He was succeeded in August 1953 by Mr B. R. Rae, MBE. Mr Rae had been closely associated with the RNZAF for many years, first as inspecting accountant at the Air Department from 1939 to 1940 and then as a serving officer (in the rank of squadron leader) and financial adviser for New Zealand's participation in the Commonwealth Air Training Plan. He had served as New Zealand Trade Commissioner in Montreal, Canada, for two years before his appointment as Assistant Air Secretary in March 1953.

Air Vice-Marshal Merton was a strong advocate for better service conditions for RNZAF members and many of the reforms he fought for were achieved. He encouraged the formation of an Air League of New Zealand in which Sir Matthew Oram, MBE, a retired Speaker of the House of Representatives, and Group Captain C. C. N. Johnston, MBE, a retired Director of Manning, were prime movers, to help promote the future of military and civil aviation in New Zealand. He firmly believed in the need for an air-minded public and saw air power as the dominant factor in any future armed conflict.

Before he returned to the United Kingdom Merton gave his own sword to the RNZAF to be awarded to the graduate at No. 1 Flying Training School and the Air Navigation and Signals School 'whose competence as a pilot or navigator is matched by his potential qualities as an officer, emphasis being placed on leadership'. In making the gift he established a tradition for the RNZAF in keeping with those long fostered in the Royal Air Force at institutions such as the RAF College, Cranwell.

First Goodwill Mission to the United States

In 1956 the Royal New Zealand Air Force sent its first goodwill mission to the United States, responding to an invitation extended by General Nathan F. Twining, Chief of Staff of the United States Air Force. The party of 18 officers and senior NCOs was headed by the Chief of Air Staff-designate, Air Vice-Marshal C. E. Kay, who had been named to succeed Air Vice-Marshal Merton on his return from the United States. The mission reciprocated official visits to New Zealand by senior USAF officers in 1954 and 1955 named Operation Handclasp and Operation Handclasp II, and was inevitably codenamed Operation Handclasp III. The mission spent three weeks in visiting American military bases as guests of the United States Government and travelled in a Hastings aircraft of No. 40 Squadron, captained by Squadron Leader K. B. Smith.

For the 21st anniversary of the RNZAF, in 1958, No. 75 Squadron mounted a 'Jetobatics' team of four Vampires.

Dakota NZ3551 over Mount Egmont.

A Sunderland flying-boat landing at the RNZAF maritime base, Lauthala Bay.

CHAPTER TWENTY

Operations in Antarctica

The RNZAF's association with Antarctic operations began in October 1955 when the United States Navy's Deep Freeze Task Force 43 was based at Wigram with eight aircraft. The station provided full technical and domestic support, but when the American aircraft were heavily loaded they used Harewood runways for take-off. Wigram, however, had the honour of being the landing point for the first aircraft which flew from McMurdo Sound to New Zealand. During this 1955–56 summer season, Squadron Leader J. R. Claydon, who had been commanding the Flying Training School at Wigram, was attached to the British Antarctic Expedition to gain experience of flying operations on the southernmost continent.

The following season the RNZAF began its own operations as an element of the Commonwealth Trans-Antarctic Expedition of 1955-58, and this gave New Zealand the opportunity to establish a station in Antarctica known as Scott Base.

In addition to various scientific studies, the primary objective of the expedition was for the British party, led by Dr Vivian Fuchs, to cross Antarctica by tracked vehicles from Shackleton Base in the Weddell Sea to Scott Base in the Ross Sea, via the South Pole. The task of the New Zealand party, led by Sir Edmund Hillary of Mount Everest fame, was to establish a series of food and fuel depots along the trans-polar route on the 8000–10,000 foot plateau towards the South Pole, for use by the crossing party.

The RAF and the RNZAF each provided an Antarctic Flight for attachment to the respective parties. The RNZAF Antarctic Flight was formed at Wigram on 1 May 1956 with Squadron Leader Claydon in command. The other members of the flight were Flying Officer W. J. Cranfield, second pilot, and Sergeant W. L. Tarr, an engine fitter. Squadron Leader Claydon, in addition to his Antarctic experience, had flown with 14 Squadron in the Second World War and with the Occupation Forces in Japan. Flying Officer Cranfield had been on the staff of the FTS at Wigram as a flying instructor. Sergeant Tarr had served with 5 Squadron at Lauthala Bay, with 14 Squadron in Cyprus and with 75 Squadron at Ohakea.

The flight was equipped with two aircraft, a de Havilland Beaver and an Auster Mark 7. The Beaver, which carried five passengers or half a ton of freight, operated with ski wheels or skis. Special equipment for Antarctic operations included engine oil dilution, a Bendix Polar Path gyro compass, an astro compass, a high frequency

Squadron Leader J. R. Claydon.

The Beaver on the ice.

The Trans-Antarctic Expedition Auster.

radio transmitter and receiver, SARAH (a search and rescue homing aid for use with the trail parties), bomb racks for carrying dog sledges and for supply dropping and a fixed F24 camera for oblique aerial photography. The Auster could be fitted with wheels, skis or floats.

Both pilots first completed courses in astro and polar navigation and aerial photography and obtained Beaver flying experience with a commercial aviation company and Auster floatplane experience at Hobsonville.

Sir Edmund Hillary was given the task of establishing an expedition training camp on the Tasman Glacier at Mount Cook and during the camp the Beaver was used on wheel skis to transport personnel, dog sledges, huskies and supplies. The Auster was also used but suffered a mishap and overturned when experimenting with a new type of wheel ski. It was repaired in time for the departure south.

On 17 December 1956 the Duke of Edinburgh, who was visiting New Zealand, inspected the expedition ship, HMNZS *Endeavour*, and met the RNZAF party. Later that night HMNZS *Endeavour* sailed from Lyttelton, but as the ship's stern swung away from the wharf the Auster in its cradle collided with a neighbouring cargo ship and the aircraft's starboard wing was severely damaged. Members of the flight were dismayed, as the Auster was intended to provide vital ice reconnaissance, so necessary for the expedition ship when passing through the pack ice. RNZAF technicians were sent by road to Port Chalmers, the next port of call for *Endeavour*, where the wing was removed. When repaired it was transported to McMurdo Sound by a Deep Freeze aircraft and fitted to the Auster in late January, several weeks after the arrival of the RNZAF Flight. The Auster was flown as a floatplane before being converted to skis.

The primary role of the RNZAF Flight was aerial reconnaissance of the Victoria Land glaciers (to determine the most suitable sledging route to the Polar Plateau from Scott Base), the air supply of depots between Scott Base and the South Pole, together with photographic and liaison flights and any emergency or special missions required by the expedition leader. Initially, this work was all done by the Beaver.

Although the normal flying season in the Antarctic is between October and February, it was decided to fly through the winter night to determine the feasibility of maintaining aircraft in a serviceable state — and of flying — in extremely low temperatures. Although the Auster was twice completely buried during severe blizzards it was flown regularly, using old-fashioned kerosene lanterns for a flarepath, these being the only type which could survive the sub-zero conditions for any length of time. Whiteouts and other flying hazards were numerous but a maximum effort was mounted during suitable weather conditions when supply flights were continued around the clock in the 24-hour summer sunshine.

The ski-equipped Beaver over Scott Base.

On 23 December 1957 Claydon and Cranfield flew in a US Navy Neptune to the South Pole and the RNZAF ensign planted there became the first New Zealand flag to fly at the Pole and the first British flag since Robert Falcon Scott's second Antarctic expedition of 1910–12.

The field parties with their huskies and sledges were flown back to base in January 1958. The final task was to fly Sir Edmund Hillary to Depot 700 to guide the British crossing party through the crevassed areas back to Scott Base. After waiting vainly over several days for the weather to clear, Claydon decided to attempt the flight on 9 February, following a new route behind the Victoria

The Auster visits a field party.

Mountain range. A rather hazardous flight of more than five hours brought the heavily laden Beaver to the Depot at 11 p.m. where it was picketed down on the plateau before returning to base the next day.

At the end of the 1957–58 season the Beaver was dismantled and placed in storage. In the 15 months it had spent in the Antarctic this aircraft had flown 266 hours, representing a total of 31,300 miles, while the Auster had flown 34 hours or approximately 3400 miles. The flight returned to New Zealand in *Endeavour*, with the Auster aboard in March 1958. The Beaver, still dismantled, returned to New Zealand in a USAF Globemaster in October that year.

Squadron Leader Claydon was well served by the other members of his team. Flying Officer Cranfield, he said afterwards, could always be relied upon to complete the most difficult flying missions, and Sergeant Tarr's ability, knowledge and experience were of inestimable value. Tarr converted a portable hut into a first-class engineering workshop and applied his rare ingenuity to almost impossible tasks. Additional support was provided by Sergeant P. H. Tate, radio mechanic, and Leading Aircraftmen I. J. Chapman, engine fitter, and A. M. Breeze, airframe fitter, who had joined the flight at Scott Base in October 1957.

'The highest praise is due to the RNZAF contingent,' said the official report of the Trans-Antarctic Expedition, presented in June 1958. 'Squadron Leader Claydon and Flying Officer Cranfield did a magnificent flying job. Sergeant Tarr and his men worked untiringly and efficiently in keeping the planes airworthy for the maximum possible time. Without the efforts of these men, routes could not have been reconnoitred so quickly, nor could the field parties have been maintained and the depots stocked.' Squadron Leader Claydon and his party were awarded the Polar Medal and Squadron Leader Claydon also received the Air Force Cross.

Following Cabinet authority for continuation of operations in Antarctica, the RNZAF Flight was reactivated in April 1959 with Squadron Leader L. C. Jeffs (Commanding Officer), Flight Lieutenant Cranfield and Flight Lieutenant P. S. Rule. This assignment, however, was shortlived. The aircraft were reassembled in Antarctica in mid December but on 15 January the Beaver crashed while returning from the Polar Plateau, near the Beardmore Glacier, during whiteout conditions. Both Squadron Leader Jeffs and Flight Lieutenant Rule escaped with minor cuts and bruises.

The US Navy tried to recover the crew by using a Dakota on skis but the terrain was too hazardous to attempt a landing. Finally, Flight Lieutenant Cranfield carried out a remarkable rescue mission in the Auster, requiring two refuelling stops en route, and in bad weather conditions brought each of the other pilots down to safety on the Ross Ice Shelf over a period of two days. For this exploit Cranfield was awarded the Air Force Cross.

There were distinctive hazards in Antarctic flying and accidents such as this had to be accepted as an operational risk, but with the loss of this aircraft New Zealand air support in the region ceased for the time being. A replacement aircraft, an Otter, was purchased in late 1960, and plans were made for deployment south again in the 1960–61 season. These were not realised, however, and although in each of the following two seasons an RNZAF pilot was attached to US Navy VXE-6 Squadron, no RNZAF aircraft flew in Antarctica until the arrival of the first Hercules there in 1965. The next RNZAF aircraft to be based there was an Iroquois helicopter which was transported to McMurdo Sound in 1985 aboard a Hercules of No. 40 Squadron to serve alongside VXE-6 Squadron in support of New Zealand and United States scientific field programmes.

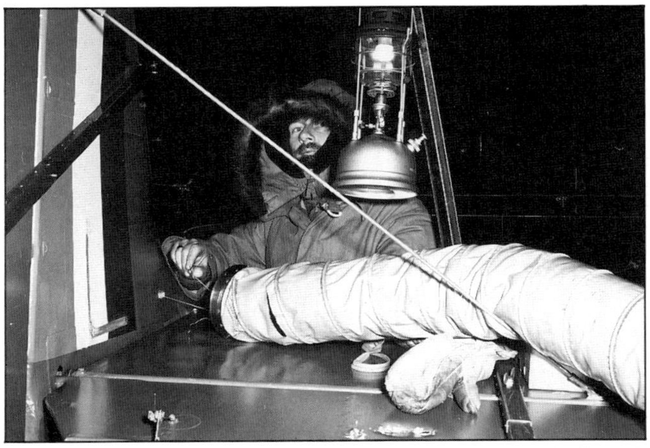

Sergeant Tarr repairing torn fabric on the Auster with the aid of a hot-air blower. *J. R. Claydon Collection*

Opposite page: Squadron Leader Claydon (right) and Flying Officer Cranfield alongside the RNZAF Ensign which they hoisted at the South Pole below the United Nations flag, permanently displayed at 10,000 feet, on 23 December 1957. The pole lies within a ring of 191 fuel drums. The RNZAF Ensign was the first British flag to be flown at the South Pole since Scott's 1912 visit. *J. R. Claydon Collection*

Opposite page: The Beaver at Depot 700 with the Snocat of the British crossing party. *J. R. Claydon Collection*

The Antarctic Flight Auster, on floats, takes off in McMurdo Sound with Mount Erebus in the background. *J. R. Claydon Collection*

The Antarctic Flight Auster is dug out after being buried during a blizzard. *J. R. Claydon Collection*

CHAPTER TWENTY-ONE

The Malayan Emergency and Confrontation with Indonesia

When the New Zealand Government, responding to Britain's 1954 request for a contribution to the Commonwealth Strategic Reserve in the Far East, decided to transfer No. 14 (Fighter) Squadron from Cyprus to Singapore, it also proposed that No. 41 (Transport) Squadron be redeployed to the Malayan theatre. So in May 1955, this squadron, commanded by Squadron Leader R. A. Manners, whose previous service with the squadron's A Flight had been recognised with the award of the MBE, returned to Changi, equipped with four Bristol Freighters, for operations against the Communist terrorist (CT) forces in Malaya. As the squadron members settled in, No. 14 Squadron was adjusting to a new living and working pattern at RAF Station Tengah, on the other side of Singapore Island. The Malayan Emergency, declared in June 1948, was to continue until July 1960.

Operating as an integral part of the RAF's Far East Air Force, the Bristol Freighters of No. 41 Squadron were used mainly for route flying, at which the crews were most proficient, and in supply dropping to ground forces in the jungle, a role for which the Bristol Freighter was ideally suited. On such missions two or more British Army dispatchers were added to the crew of three. Supply dropping was demanding work, calling for concentration and precise flying. In the first two years in the theatre much of the supply dropping was handled by Flight Lieutenants G. G. Bayliss, DFC, and J. N. Trolove.

Route flying took the Freighters into almost every Asian and South-East Asian country, from Nepal to Japan. Other tasks undertaken by the squadron included general courier flights and support for exercises mounted by the South-East Asia Treaty Organisation (SEATO). The squadron was also frequently called upon to fly mercy missions, as air ambulance flights were termed. On one such mission, in April 1956, Squadron Leader Manners flew to Phuket in Thailand to uplift a man with a knife wound in his skull which had exposed his brain. The patient was flown to Penang, for hospital treatment, with altitude maintained at 500 feet throughout the flight to avoid pressure on the brain.

In June 1956, when Britain was about to explode an atomic device at Monte Bello Island,* off the north-west coast of Australia, and had Canberra bombers positioned at Perth to monitor the test, Squadron Leader Manners flew a consignment of specially made canopy covers from Singapore to Perth for use by the Canberras in their mission. On the return flight three days later, he had to fly the aircraft below 2000 feet throughout the eight and a half hour journey because of the atomic explosion.

* The first British atomic weapon was exploded at Monte Bello on 3 October 1952.

Squadron Leader R. A. Manners, MBE, officer commanding No. 41 Squadron, photographed at Whenuapai with the squadron training officer, Flight Lieutenant J. N. Trolove, AFC (right) and Flight Lieutenant J. A. Houlton, DFC (centre), before the squadron's move to Singapore.

At the end of a busy tour of duty in the Far East Manners handed over command of 41 Squadron in November 1956 to Squadron Leader A. S. Tie. On 10 December a Bristol Freighter took off on a mission to drop supplies to ground forces in the Cameron Highlands. The aircraft was commanded by Squadron Leader Tie, who had on board, besides his two RNZAF crew members, four British Army air dispatchers and two cameramen of the Malayan Film Unit. Flying in bad weather, the Freighter crashed about noon into a steep ridge covered in dense primary jungle, at an altitude of about 4200 feet. The wreckage was later sighted by an RAF spotter aircraft but because of the appalling weather two days elapsed before an Auster could guide rescue patrols to the area. The only survivor, one of the British Army dispatchers, equipped with rations from the supplies on board, walked out; he was found by Malay troops about 12 days later. Squadron Leader G. H. S. Tosland arrived in January 1957 to replace Squadron Leader Tie.

At Tengah, No. 14 (Fighter) Squadron again flew aircraft hired from the RAF. The establishment for the squadron provided for 16 de Havilland Venom FB.1 fighter-bombers and two de Havilland T.11 two-seat trainers, but few Venoms were immediately available and Squadron Leader N. H. Bright and his pilots had to make do with an assortment of Vampire Mark 9s, with which

From 1955 through to the 1970s many Commonwealth soldiers in South-East Asia's jungles had this view of their supplies being dropped by a Bristol Freighter of No. 41 Squadron.

A Venom of No. 14 Squadron in a rocket attack on a Communist terrorist position in the Malayan jungle.

they were quite at home, and two-seat Mark IIs.

The squadron served alongside the RAF's No. 60 Squadron, which also flew Venoms, and the Royal Australian Air Force's No. 1 Squadron, equipped with Avro Lincoln heavy bombers, under the command of Headquarters, Far East Air Force (FEAF). These squadrons flew in support of British Commonwealth ground forces combating Communist terrorist activity in the region. These ground forces included, in 1956 and 1957, the newly raised New Zealand Special Air Service (SAS) Squadron and, later, units of the New Zealand Regiment.

Army troops were deployed as intelligence reports disclosed the locations of CT forces in the Malayan jungle and air support was mounted as required by Headquarters, FEAF. The preplanned bombing, strafing and rocket attacks — of which No. 14 Squadron contributed its share, with other Commonwealth squadrons — were labelled Firedog operations. The immediate response operations, codenamed Smash Hits, were undertaken by duty air and ground crews who were available 24 hours a day.

Flight Lieutenant S. McIntyre who, after his tour of duty in Cyprus had served with 75 Squadron at Ohakea, had been posted back to 14 Squadron as one of its two flight commanders and he had assumed temporary command of the squadron while Squadron Leader Bright and Flight Lieutenant C. J. Berryman, the senior flight commander, were in transit between Cyprus and Singapore. It was Flight Lieutenant McIntyre who launched the first 14 Squadron offensive when he led a formation of five Vampires in a strike against terrorist positions on 1 May 1955. He flew the first Venom on a Firedog mission on 6 May, the first of the squadron's new aircraft, Venom WR366, having been recently delivered. (There had been keen competition between the New Zealanders and No. 60 Squadron to be the first squadron to use this type of aircraft on operations.) It was to be March 1956, however, before No. 14 Squadron finally gained its full complement of 16 Venoms.

In a typical Firedog a section of four Venoms would make a strike against jungle positions, marked by RAF Austers, each Venom carrying either two 1000 lb bombs or eight 60 lb rockets and always with 600 rounds of 20 mm cannon ammunition. Often the New Zealanders flew as part of a FEAF strike force such as the large-scale mission on 18 June 1956 when 22 Venoms, eight Canberra bombers and six Lincoln bombers raided a key terrorist camp north of Kluang. That month the squadron flew 10 Firedogs and three Smash Hits.

The Venom fighter-bomber, although it closely resembled the Vampire in appearance, was a faster aircraft with a higher rate of climb because of the increased thrust provided by its DH Ghost engine — a 4850 lb thrust compared with the DH Goblin's 3350 lb — and the redesigned wing took full advantage of this. The pilots, whose average age was 21, liked the aircraft which was pleasant to fly. It was not well suited to the tropics, however, and the aircraft was eventually 'lifed' at 750 hours. During one 12-month period the squadron scrapped 19 aircraft, only five of them because of accidents.

When 14 Squadron took over their Venoms the aircraft were distinguished from those of 60 Squadron by a distinctive red and white chequered pattern painted on the twin rudders. Later the rudders were painted black with a white kiwi superimposed, and these markings were retained for the remainder of the squadron's service in the Far East.

By late 1955 the squadron had no further need of its single-seat Vampire 9s although the last one remained on strength until 3 February when Flight Lieutenant McIntyre delivered it to Seletar. Late the following month he took delivery of the first of the latest ejection seat fitted Vampire T.11s, going to Mingaladon in Burma with Flying Officer K. Milne to collect it. From the outset T.11s were flown on anti-terrorist operations, at least until the squadron had its full complement of Venoms, but their primary role was for instrument flying and flight checks.

As soon as sufficient Venom aircraft became available, and operational flying commitments permitted, Squadron Leader Bright reformed his four-man formation aerobatic team, which had performed with such distinction in Cyprus. The Venom aerobatic team gave its first public exhibition at Sembawang before SEATO conference delegates on 27 June 1956, and also performed before the Commander-in-Chief, Far East Air Force, Air Marshal the Earl of Bandon, the New Zealand Chief of Air Staff, Air Vice-Marshal C. E. Kay, and the New Zealand High Commissioner in South-East Asia, Mr Foss Shanahan, on 17 August that year.

During Squadron Leader Bright's term as officer commanding, the squadron had a few accidents, probably the most memorable being Pilot Officer Ian Ferguson's crash landing into rubber trees at Tengah while practising touch and go landings on 4 April 1956. When Squadron Leader Bright arrived at the scene he was amazed to see Pilot Officer Ferguson standing among a group of spectators watching his dismembered aircraft burning fiercely. Some time later a medical officer came close to being decapitated when he went to examine the wrecked aircraft; he was about to peer into the cockpit when the ejection seat exploded upwards with startling suddenness.

On 20 January 1956 a Vampire T.11 was lost when Flying Officer C. W. Rudd was forced to ditch his aircraft after an engine failure off the tip of Malaya. He put the aircraft down in about 5 feet of water, but as it began to sink he found his radio was able to transmit and he advised base that both occupants were safe and uninjured. Malayans brought them ashore in a canoe and the RAF returned them to their unit by helicopter.

Squadron Leader Bright handed over command to Squadron Leader A. F. Tucker on 24 September 1956. Flight Lieutenant McIntyre served as the senior flight commander under Squadron Leader Tucker and Flight Lieutenant E. C. Arundel replaced Flight Lieutenant Berryman, whose tour of duty had expired. After his return to New Zealand Flight Lieutenant McIntyre was awarded the DFC.

Although No. 14 Squadron's anti-terrorist operations did not result in any casualties to squadron members, there were several dramatic incidents in 1957. On 29 April Flight Lieutenant M. F. McD. Palmer, who had succeeded Flight Lieutenant Arundel as a flight commander, was leading a flight of four Venoms on a low-level mission over Johore Bahru when the engine failed. He instantly sought greater height with what momentum remained and attempted to relight but without success. He then used his ejection seat, only too well aware that at less than 1000 feet he had barely enough altitude. The parachute opened and almost at the same instant the pilot hit the water in the Straits of Johore. Flight Lieutenant Palmer inflated his rubber dinghy, but a police launch — which the falling aircraft narrowly missed — was alongside him before he could make use of it and he was taken on board. This was the first time that a pilot ejection seat had been used to escape from an aircraft in RNZAF service.

Several weeks later, on 3 July, 22-year-old Flying Officer G. C. Hubbard was flying over the Malayan jungle at 20,000 feet when his Venom went into a spin. Unable to regain control, he ejected at 12,000 feet and landed safely three-quarters of a mile from the spot where his Venom had crashed. Because of language problems and the absence of communication facilities, his whereabouts remained unknown for at least four hours before

Venoms return to Tengah, mission completed.

A Venom rocket strike.

he was found by Malayan policemen and returned to his unit.

Squadron Leader Tucker led 14 Squadron until May 1958 when the Venoms were handed back to the RAF after the Canberra bomber was introduced into the RNZAF. Following a deployment in April to Don Muang air base in Thailand, where the squadron exercised with USAF and Royal Thai Air Force squadrons, Squadron Leader Tucker staged a farewell flypast over Singapore on 16 May, 10 Venoms and three Vampire T.11s flying in a figure 14 formation. A display of aerobatics was given over the Singapore waterfront where, more than 16 years earlier, the gallant survivors of 488 Squadron had been shipped to Batavia. When 14 Squadron ceased its Venom operations it had dropped 500,000 lb of bombs and fired 1523 rockets on to terrorist positions, and although there had been aircraft losses the squadron had suffered no fatalities, a record of which it could justly be proud. Squadron Leader Tucker was awarded the DFC on 28 October 1958, in recognition of his leadership of 14 Squadron during the Malayan Emergency.

The introduction of the English Electric Canberra bomber as the RNZAF's front-line strike aircraft and the first few years of Canberra operations happened to coincide with the formation of the independent Federation of Malaysia and the subsequent border conflicts with Indonesia, the latter known in political and military circles as 'the Confrontation'.

On 16 September 1963 the Federation of Malaya* united with Singapore and the two former British colonies Sarawak and Sabah (or North Borneo) to form a new federation known as Malaysia. On 9 August 1965 Singapore withdrew from the federation, leaving 13 remaining states, most of them in Peninsular Malaysia, with two, Sarawak and Sabah, on the northern coast of the island of Borneo. The trouble with Indonesian infiltration reached critical proportions in late 1964 when both 41 Squadron and 14 Squadron, the latter equipped with Canberra B(I).12 bombers, assisted Britain's Far East Air Force in a successful bid to contain Indonesian territorial ambitions.

* The Malay States had ceased to be British colonies on 31 August 1957 when they became the Federation of Malaya.

RNZAF Canberra operations commenced in July 1958 when 75 Squadron crews, under Squadron Leader G. R. B. Highet, DFC, AFC, having completed conversion training in England, ferried their hired RAF Canberra B.2 light bombers from the United Kingdom to Singapore. No. 45 Squadron, RAF, flew B.2s to Singapore at the same time and the two squadrons were formed into a bomber wing at Tengah, to serve as part of No. 224 Group, Far East Air Force, in the final stages of the Malayan Emergency.

At Tengah 75 Squadron took over the buildings and hardstanding area recently vacated by No. 14 Squadron. The squadron had its own first-line and second-line servicing teams under Flight Lieutenant R. F. Ward and also provided a proportionate number of tradesmen in base servicing facilities and in non-technical sections and messes of the Tengah base.

Engaged for a time on anti-terrorist bombing operations, 75 Squadron trained regularly with 45 Squadron and with RAAF B.20 Canberra squadrons based at Butterworth. Several exchange visits with USAF B-57 squadrons (the B-57 was a Canberra built under licence in America) were hosted by the squadron under the codename Exercise Joss Stick and these visits were reciprocated when 75 Squadron detachments flew to the Philippines, Thailand and Okinawa. Joss Stick 10 was the last such visit, to USAF Base Kadena on Okinawa in August 1961. 75 Squadron aircraft and crews, together with those of No. 41 Squadron, represented New Zealand in the annual Manila Air Show in 1959 and again in 1960 and 1961.

On a regular basis, single Canberra aircraft drawn from No. 75 Squadron, 45 Squadron and the RAAF squadrons carried out long-range navigation exercises known as 'lone ranger' flights to New Zealand, Australia and Hong Kong.

Towards the end of 1958, the first of the RNZAF's new B(I).12 Canberra bombers were in production at Samlesbury, near Preston, England, having been ordered in 1957 to re-equip No. 14 Squadron. In January the following year, 32 officers, NCOs and airmen who would form the nucleus of the new No. 14 Squadron left Whenuapai in a Hastings of No. 40 Squadron to undergo air and ground

A Canberra B.12 in the camouflage adopted for operations in Malaysia during the Confrontation.

The Duke of Edinburgh inspects airmen of No. 75 Squadron in February 1959, accompanied by the squadron commander, Squadron Leader G. R. B. Highet.

Aircrew members of No. 75 Squadron at Tengah in 1960. From left: Flight Lieutenants P. J. Heares, R. V. Smith and M. D. Peacock, Flying Officer S. E. Gillingham, and Flight Lieutenants I. M. Gillard and R. C. Macfarlane.

training in the United Kingdom. Twenty-five of this party were engine, airframe, electrical, armament and instrument fitters who, led by Squadron Leader C. M. Crawford, MBE, were to train at various British factories and RAF stations.

By way of transition from the fighter/ground-attack role to the light bomber role, No. 14 Squadron was redesignated the Jet Holding Flight of the Bomber Operational Conversion Unit at Ohakea on 23 November 1959, and it was reactivated as a light bomber/interdictor squadron on 1 March 1960 under Squadron Leader H. G. Moss.

Designed for the low-level interdiction role, the Canberra B(I).12 was an aircraft of great versatility; it had an airframe specially strengthened for manoeuvring and weapons delivery at low level and it could carry a wide range of tactical weapons over considerable distances. With the thrust from its Rolls Royce Avon 109 engines it had the power to exceed its limiting indicated airspeed of 450 knots in level flight.

The RNZAF Canberra purchase included 11 B(I).12s and two T.13 trainers. Powered by two Rolls Royce Avon 1 engines, the T.13 could carry a crew of three, but the B(I).12s were configured to carry only a pilot and a navigator. The initial B(I).12 deliveries were made in late 1959. Because delivery of the T.13s was delayed for the development and fitting of pilot ejection seats (the RAF's T.4, the model on which the T.13 trainer was based, had an ejection seat for the navigator only) the RNZAF hired two T.4s from the RAF so that conversion training of 14 Squadron crews could proceed. The T.4s were uplifted from RAAF Edinburgh, near Adelaide, Australia, in February 1960 by 14 Squadron crews. (They were ferried back to the RAF early in 1961 after the T.13s had arrived.)

In January 1961 Squadron Leader Moss succeeded Squadron Leader Highet at Tengah as 75 Squadron's commanding officer. Early the following year the New Zealand Government's Review of Defence Policy brought about the recall of No. 75 Squadron from Singapore as an economic measure designed to conserve overseas funds — the Malayan Emergency having officially ended in July 1960 — and the hired B.2s were returned to the RAF, being ferried to the United Kingdom by 75 Squadron crews.

Accompanied by Wing Commander A. F. Tucker, DFC, head of the New Zealand Defence Liaison Staff in Singapore, Squadron Leader Moss flew to Kuala Lumpur on 23 February 1962 when the Prime Minister, Tun Abdul Razak, attended by the New Zealand High Commissioner, Mr C. M. Bennett, thanked him for the fine work that 75 Squadron had done in his country over the past four years.

Squadron personnel were progressively returned to New Zealand and by the end of March 1962 all that remained of their presence in South-East Asia was a large measure of goodwill and a first-class reputation with the Far East Air Force and SEATO allies. With the premature repatriation of squadron members, the Air Force was at pains to absorb them into other units with a minimum of personal hardship to the men and their families.

No. 14 Squadron, with its Canberra B(I).12s, then became a New Zealand-based component of the Commonwealth Strategic Reserve and was required to make frequent deployments to South-East Asia, visits code-named Exercise Vanguard. The Vanguard series took the New Zealand crews to Tengah, and to the RAAF base at Butterworth. Lone ranger flights during these deployments allowed crews to fly even further afield, to Kai Tak in Hong Kong and to Clark Air Force Base in the Philippines.

In September 1964 No. 14 Squadron, then commanded by Squadron Leader G. Wallingford,* flew to Singapore on what was publicly known as Vanguard VI, another routine, short-term deployment. The months preceding September had seen an upsurge of Indonesian activity against the new Federation of Malaysia, air incursions across the Kalimantan border and even raiding parties entering West Malaysia. An air defence identification zone had been established in February that year along the border between East Malaysia and Indonesian territory. During September the Indonesians launched air attacks against villages on the East Malaysian side of the border, and after the dropping of paratroops in West Malaysia (in Johore, across the causeway from Singapore) on 9 September the Far East Air Force was placed on full alert. The uneasy relationship between Indonesia and Malaysia stemming from these border conflicts resulted in 14 Squadron remaining at Singapore as part of the Far East Air Force.

Under the guise of another Vanguard Exercise, No. 14 Squadron had, in fact, been deployed to Singapore in the hope that its presence there, along with other Commonwealth units, would have a deterrent effect on mounting aggression in the area. Before leaving New Zealand the six B(I).12s had been quickly painted in camouflage and had been fitted with UHF radio. The decision to deploy the squadron was taken on 14 September, the airmen were told the following day, and the first four aircraft took off from Ohakea on 16 September, 37 hours after the order was received. The squadron flew to Singapore via RAAF Edinburgh, Perth and Cocos Island and then north around Sumatra to avoid Indonesian airspace. The supporting Hastings aircraft of No. 40 Squadron flew the same route many times during September and October.

The New Zealand Canberras were armed at once. During the previous week, to counter the Indonesian threat, the Far East Air Force had bolstered the strength of its strike force to 96 aircraft with the arrival of a Vulcan bomber detachment, three Canberra strike squadrons, one Canberra photo reconnaissance squadron and a Javelin all-weather interceptor squadron. The aircraft were dispersed, vital points on the airfield were sandbagged and guarded and anti-aircraft gun positions were established. The Javelin fighters stood on quick alert at the end of the

* A son of Air Commodore S. Wallingford, Squadron Leader Wallingford had entered the RNZAF in 1952 as a cadet under the scheme whereby a limited number of New Zealanders could train at RAF College Cranwell. He distinguished himself there by winning the Sword of Honour as the best all-round cadet in 1954.

Canberras starting up.

runway. Only days before, the Indonesians had made incursions with paratroop drops 100 miles to the north in the Labis area and Tengah stood ready in case Indonesia decided to attack the base using its Russian-built Il-28 Beagle bombers and MiG fighters. Indonesia also had a P-51 Mustang squadron uncomfortably close on the island of Tandjung Pandan, just south of Singapore.

Tension built when the aircraft carrier HMS *Victorious* sailed through the narrow Sunda Straits between Java and Sumatra en route to Singapore from Perth. All Tengah aircrew were brought to heightened readiness in crew rooms but the Indonesians were not tempted to react, if indeed they knew that the ship was there. HMS *Victorious* was operating with one boiler out, and her capability to launch her Sea Vixen and Scimitar aircraft was seriously diminished.

For the next few months No. 14 Squadron refined its tactical formations and adapted to a level rocket release from 50 feet above the ground. The crews practised, ever mindful of the Indonesian President's vow that he would crush Malaysia by the time the cock crowed on New Year's Day, but January 1965 came and went without any such disaster.

Bombing up Canberras at Ohakea.

An Avon engine change.

Mainstay of the RNZAF Strike Wing for more than a decade, English Electric Canberra bombers, at Ohakea.

NZAF tradesmen at work servicing a DH Devon navigation trainer at Wigram.

A Bristol Freighter of No. 41 Squadron over Nui Dat, South Vietnam, bringing supplies and replacement personnel for the New Zealand Army troops based there.

RNZAF Headquarters Malaysia

When the six Canberras and the squadron's 100-strong ground component left for Singapore, one of the supporting Hastings also carried Wing Commander D. E. Jamieson, who had been entrusted with the move of the squadron to Singapore and national command of both 14 and 41 Squadrons in the theatre. His new directive as Force Commander also required him to establish and command an RNZAF Headquarters Malaysia. In this new appointment, from offices close to 224 Group Headquarters, he exercised administrative command over the two New Zealand units, which for operational purposes still came under FEAF command. Wing Commander Jamieson was much concerned in the months ahead with the highly charged political atmosphere; it was during this period that Singapore withdrew from the newly formed federation and became an independent nation.

Wing Commander Jamieson had two of his staff officers, Squadron Leader B. D. Gordon and Flight Lieutenant B. E. Gilliver, attached to 224 Group for liaison duties. Other RNZAF officers in the theatre included Wing Commander W. G. Hughes, who was serving at Kuala Lumpur as assistant defence liaison officer in the office of the High Commissioner for New Zealand in Malaysia, and Flight Lieutenant J. C. Buckmaster, AFC, who was serving on exchange duty with the RAF, flying Whirlwind helicopters from Seletar and on active operations in East Malaysia. Air Commodore S. G. Quill, OBE, DFC, was also at 224 Group, serving on exchange duty as the Group's Senior Air Staff Officer.

Wing Commander D. E. Jamieson who commanded RNZAF Headquarters Malaysia during the Confrontation with Indonesia.

No. 41 Squadron Operations

Before May 1962 No. 41 Squadron at Changi had been engaged primarily in medium-range route transport operations as part of the Commonwealth Strategic Reserve and under the operational control of Headquarters, FEAF. In May that year a developing threat to Thailand's borders by pro-Communist forces provoked a SEATO reaction and the squadron was dispatched to Thailand to assist in maintaining security in that area. The Bristol Freighters were based at Korat, in Northern Thailand, a garrison town for almost 1000 years of Asian strife. At first the quarters were most primitive but gradually the initiative for which New Zealanders are renowned improved their lot. Between May and December the squadron worked closely with the United States forces and was controlled operationally by the Commander, Joint Task Force 116, Lieutenant-General James Richardson. For this operation, codenamed Scorpion, the RNZAF detachment, under Squadron Leader B. A. Wood, was placed on active service.

In December the squadron was withdrawn to Singapore but a new crisis had arisen, and the squadron was immediately committed to Operation Borneo Territories, being placed on active service for the second time in seven months. In little more than a week of the flying-in phase

Bristol Freighters of No. 41 Squadron operated regularly throughout Thailand in the 1960s.

the Bristol Freighters had completed 230 hours of intensive around-the-clock missions. This operation, in support of the British forces fighting a rebellion in Brunei, was brief but exhausting.

By January 1963 41 Squadron had a detachment of two Freighters back at Korat in response to a request to the New Zealand Government for a contribution to the programme known as SLAT — Special Logistics Aid to Thailand. This detachment was to remain at Korat for more than two years on this project. In that time the two aircraft flew 647 courier flights, carrying 3,500,000 lb of freight, 92,000 lb of mail and 15,877 passengers. In 25 months only five courier flights were uncompleted, two because of weather, two on account of aircraft mechanical failures and one owing to pilot illness. For the 41 Squadron detachment, duty in Thailand was interesting but arduous and made difficult by the fact that Korat was a forward base, far removed from the squadron's headquarters at Changi.

When the last Freighter lifted off Korat's airfield on 20 February 1965, signifying the ending of New Zealand's support for SLAT, the crew was carrying a carved black kiwi mounted on a wooden base, a gesture of appreciation from the Commandant of the Royal Thai Air Force, Air Vice-Marshal Kamron Lee Lasiri, who had told the squadron detachment before they left his country: 'They call you Number Forty One; to us you are Number One.'

Squadron Leader R. I. F. Garrett succeeded Squadron Leader Wood as squadron commander in October 1963. During the increased activity that the confrontation period brought to the units serving at Singapore most of the work was of a routine nature but there were some tense moments. As no one knew when the Indonesians might stage an attack, the squadron had to maintain a constant state of readiness as well as ensuring security for the servicemen's families at Changi.

In September 1965 Squadron Leader Garrett handed over command to Squadron Leader I. A. Hutchins. On 13 October 1965 a Bristol Freighter captained by Flight Lieutenant N. J. S. Rodger, took off from Kuching on a supply dropping mission which required the crew to fly close to the Indonesian border. The crew included Flight Lieutenant G. Thompson, co-pilot, Flight Lieutenant N. J. Barr, navigator, Flight Sergeant M. Southam, signaller, and dispatchers from the British Army. On its approach run to the dropping zone the aircraft was flying just below 500 feet when it inadvertently passed over a strip of Indonesian territory. The Bristol was struck by a burst of machine-gun fire, about 12 bullets penetrating the fuselage. Flight Lieutenant Rodger cancelled the supply drop and returned to base. The subsequent board of inquiry found that although there had been a breach of flying regulations in passing over the border, operational necessity had been a factor leading to the incident.

14 Squadron's Tour of Duty Extended

The New Zealand Government announced in November 1964 when Vanguard VI had been under way for two months that 14 Squadron would remain in the Far East until March 1965 because of the continuing tension in the area. The period was later extended until October 1966 and Wing Commander Jamieson's last major task as Force Commander was to plan the withdrawal of 14 Squadron to New Zealand. This time the airlift was undertaken by the new C-130 Hercules transports of No. 40 Squadron* commanded by Wing Commander B. A. Wood. When 14 Squadron returned to Ohakea in November 1966, the New Zealanders had neither dropped a bomb nor fired a rocket in anger during the Confrontation. The crews had, however, gained valuable experience in an operational atmosphere, as had the ground component of the squadron. And they all had the satisfying knowledge that their presence and combat readiness had been a factor in FEAF's ability to restrain Indonesian aggression. They had witnessed at first hand a classic example of the deterrent effect of air power.

The Canberras remained in RNZAF service for another three or so years. Lone ranger navigation exercises took crews to Fiji, and there were further regular Vanguard deployments to Tengah and Butterworth. In 1970, with the advent of Skyhawks, the Canberras were phased out of RNZAF service.

In 12 years of Canberra operations the RNZAF lost three aircraft. NZ6101 was written off on 2 November 1960 when the aircraft lost power during a practice single-engine overshoot at Christchurch International Airport and was force landed just outside the airport boundary. The crew escaped serious injury. Canberra NZ6104, crewed by Flight Lieutenant J. W. Southgate and Flying Officer R. B. Thomson, crashed in the sea 30 miles east of Singapore on 13 November 1964. There were no survivors. One of the hired RAF Canberra B.2s of 75 Squadron was lost in 1961 on a night mission over the Cameron Highlands when Flight Lieutenant P. G. Bevan lost control of his aircraft in violent turbulence. Bevan ejected successfully (and walked out of the jungle four days later) but his navigator, Flight Lieutenant D. L. Finn, died in the crash.

Opposite page: Back from their training in the United States, RNZAF Hercules transport crews await an official welcome from the Prime Minister, the Rt Hon. Keith Holyoake, at Wellington Airport in 1965. One of the new Hercules aircraft stands in the background.

* See Chapter 22.

CHAPTER TWENTY-TWO

The Morrison Touch

The Government's 1961 Review of Defence Policy placed emphasis on building up the New Zealand Army, its only positive statement on strike aircraft being a reference to 75 Squadron's recall from Singapore on the grounds of economy. It proposed that the Sunderlands be retained in service but 'in progressively reduced numbers'. Re-equipment programmes, it said, would be needed for maritime, transport and training aircraft 'in the future'. The only mention of re-equipping the Air Force in the immediate future was a statement that three Douglas DC6 aircraft which Tasman Empire Airways no longer required would be given to the RNZAF 'as an interim measure'. This savoured of political expediency.

The Air Force's combat strength had been effectively halved by the recall of 75 Squadron from the Far East and the return of its Canberras to Great Britain. The Sunderland flying-boats were obsolescent if not obsolete as anti-submarine aircraft. The basic trainer was still the North American Harvard on which wartime pilots had trained a generation before.

When Air Vice-Marshal I. G. Morrison, CBE, assumed the leadership of the Service as Chief of the Air Staff in June 1962 the prospects for the future development of the RNZAF looked bleak indeed.

Sensing a lowering of morale within the Service generally, the new CAS saw his duty clearly. He firmly believed that the Air Force could be rebuilt and given new direction, even in the face of political apathy. He knew, too, that several thousand men and women were looking to him for action. Single-mindedly, he bent himself to the task of re-equipping the RNZAF for all its roles.

If Cochrane was the right man at the right time in 1937, so too was Morrison in the year 1962. Within a few weeks of taking office he was talking to his men on their own ground, telling them of his plans for the Service, inspiring them, rekindling optimism, giving them a new sense of purpose. He visited every RNZAF station, asking the station commander to assemble every man and woman who could be spared from a post for 40 minutes. Then he outlined his blueprint for the RNZAF.

Air Vice-Marshal I. G. Morrison, CB, CBE, the RNZAF's tenth Chief of the Air Staff. His determined leadership took the Air Force out of a trough and into a new, vital period of achievement, with first-class modern aircraft. *National Publicity Studios*

He also took the national press into his confidence and ensured that the news media were well briefed on the background to re-equipment, a move which brought some heavy guns to bear on public opinion as he started to press the RNZAF case with the Government. He was outspoken and was several times rebuked by senior politicians for expressing his personal views on such subjects as an aircraft's suitability for the RNZAF. But he steadfastly refused to put his own future before that of his Service.

Morrison declared the three basic roles of the Air Force — strike, maritime reconnaissance and air transport — to be of equal priority. To begin with, he advocated re-equipping the long-range transport squadron with

A No. 5 Squadron Orion on ocean reconnaissance.

Lockheed C-130 Hercules aircraft and seeking a maritime version of the same type to replace the Sunderlands. He envisaged 12 Hercules — seven for transport and training and five for the maritime squadron.

Further examination proved that a maritime C-130 was not the answer. An evaluation team, led by Air Commodore W. H. Stratton, DFC and Bar, Assistant Chief of the Air Staff, visited the United States and came out in favour of the Lockheed P-3 Orion, a long-range maritime patrol aircraft capable of speeds of more than 400 mph, with which the United States Navy was then equipping. Morrison agreed and a submission was drawn up recommending the purchase of Orions for the maritime role and Hercules for long-range transport. A modern combat aircraft, probably of American origin, was a major part of Morrison's plan and he also envisaged the introduction of helicopters.

On the organisational side, the Minister of Defence, the Hon. D. J. Eyre, had for some time been urging the unification of the three Armed Services with the creation of a single Defence Ministry, believing that both economies and greater efficiency would result. The 1961 Defence Review had expressed disappointment that little progress had been made towards integration and throughout 1963 and 1964 integration of the Armed Forces was introduced progressively. Mr J. K. Hunn, CMG, was appointed the first Secretary of Defence in 1963 and Rear-Admiral (later Sir) Peter Phipps, CBE, DSC and Bar, VRD, became the first Chief of the Defence Staff in the rank of vice-admiral. The following year, on 20 November, a Defence Council was constituted, and Mr J. L. Jenkins (who had succeeded Mr B. R. Rae as Air Secretary in April 1964) was appointed Deputy Secretary of Defence (Air).

Meanwhile, in May 1963 the Cabinet had approved the Air Force submission for the purchase of three Hercules aircraft and these arrived in April 1965, being flown to New Zealand by RNZAF crews. Two further Hercules were ordered, and these were delivered in January 1969. The purchase of five Orions was authorised by Cabinet in August 1964 and deliveries were completed in 1966.

Command of 40 Squadron when it re-equipped with Hercules aircraft was given to Wing Commander B. A. Wood, AFC, who had commanded 41 Squadron in Singapore and Thailand. The C-130s quickly proved their worth and undertook a remarkable amount of work in the first few years of their operations.

The first five Orions of No. 5 Squadron lined up on the tarmac at Whenuapai.

In April 1965 the Minister of Defence announced the Government's decision to buy five Bell Iroquois and six Bell 47G Sioux helicopters for Air Force and Army operations, besides two Westland Wasp anti-submarine-warfare helicopters for use on Royal New Zealand Navy Leander-class frigates.* The Sioux was for visual reconnaissance in support of the Army and basic flying training while the Iroquois was for advanced and operational training, battlefield transport in support of the Army and search and rescue.

The introduction of helicopters not only meant that a surprisingly large proportion of the Air Force's general duties officers acquired totally new skills but it also saw the creation of a tri-Service squadron whose new and exacting role would take its members to the South Pacific island territories, the Antarctic, South Vietnam, the Sinai Peninsula and on sea duty with the Royal New Zealand Navy. This distinctive new unit, based at Auckland, was initially named No. 3 (Battlefield Support) Squadron, thus reactivating an old and honoured unit which had been the first squadron to go into action in the South-West Pacific in the Second World War. (The 'Battlefield Support' appellation was formally dropped in May 1969.) Over the next 22 years No. 3 Squadron was to undertake a heavy catalogue of important commitments more diverse in nature and scope than those of any RNZAF unit before it.

By 1966 all but one of Air Vice-Marshal Morrison's objectives had been realised. The Hercules had already established themselves in long-range transport operations; the Orions, based at Whenuapai† under Wing Commander D. B. Flintoff, were justifying their claim to be the finest anti-submarine aircraft in the world; and the helicopters of No. 3 (Battlefield Support) Squadron, under Wing Commander C. L. Siegert, were proving invaluable, not only in support of the three Services but also in search and rescue. But a new strike aircraft had not yet been selected.

* Another seven Sioux and 10 Iroquois were delivered in 1970. Further Wasps kept pace with RNZN frigate additions.

† When the Orions first went into service in 1966 they did so as the Orion Flight of the Maritime Wing at RNZAF Base Auckland. The Sunderlands continued in service until March 1967. On 31 March 1967 RNZAF Station Lauthala Bay closed down and on 1 April the Orion Flight officially became No. 5 Squadron.

A Sioux helicopter on floats.

The Air Force had also completed a major reorganisation, the changes being implemented progressively. With effect from 1 September 1965, an RNZAF Operations Group had been established under an Air Officer Commanding, with headquarters at Whenuapai, embracing the Strike Wing at Ohakea, the Maritime and Transport Wings at Auckland, RNZAF Headquarters Malaysia, and several other units such as the RNZAF detachment at Lauthala Bay, the Command and Staff College at Whenuapai, No. 1 Stores Depot (Te Rapa) and No. 1 Technical Training School. An RNZAF Training Group, with headquarters at Wigram, was to be responsible for all training functions of the Air Force, both flying and ground, and it brought Wigram and Woodbourne under a single command, besides certain other South Island-based units.

Further enduring changes became effective on 1 February 1966 — RNZAF stations were disbanded as such and recreated as bases, and Hobsonville and Whenuapai were amalgamated to become RNZAF Base Auckland. In their new role, bases were required to give operational, technical and administrative support to those Operations

A Hercules of No. 40 Squadron takes off from a Fijian island airstrip.

Iroquois helicopters at RNZAF Base Auckland.

Group and Training Group* formations lodged on them. With these changes, the term 'station commander', which had been in use for almost 30 years, passed out of the Air Force vocabulary.

When Air Vice-Marshal Morrison retired in 1966, he was able to hand over to his successor an Air Force that was much better equipped and motivated than when he had assumed command in 1962. Only in respect of new strike aircraft had there been no decision.

Air Vice-Marshal C. A. Turner, CBE, gave the selection of a suitable aircraft to fit the strike role the highest priority. The original contenders included the American F-4 Phantom and F-5 Freedom Fighter, and the Australian version of the French-designed Mirage fighter. There was considerable pressure on the Air Force, and on the New Zealand Government, to buy the Mirage; the fact that this aircraft was then in service with the RAAF was seen by some as a deciding factor in its favour.

But the aircraft finally chosen was the McDonnell Douglas A-4 Skyhawk. The purchase included 10 single-seat A-4Ks and four TA-4Ks (dual-control, two-seat trainers) at a total cost of $NZ24.6 million. The Skyhawk, originally designed and built by the Douglas Aircraft Corporation at Longbeach, California, had first flown in June 1954, but when the RNZAF placed its order in June 1968 the A-4 was still a mainstay of the United States Navy and Marine Corps air attack wings. The smallest, lightest and toughest combat aircraft in the US Navy, it was highly rated for its bomb delivery accuracy and low loss rate in combat in Vietnam. It had been developed and refined over many years as a specialised highly efficient subsonic attack aircraft, and this was one reason why the RNZAF selected it for its strike role.

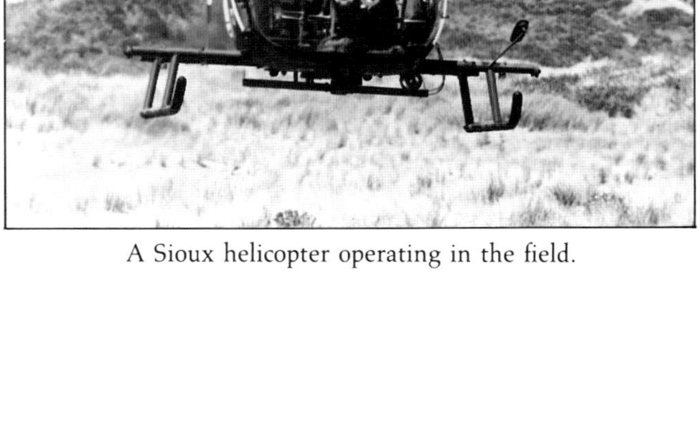

A Sioux helicopter operating in the field.

* With effect from 1 June 1972 there was further reorganisation. Training Group was renamed Support Group, and units such as the Command and Staff College, No. 1 Technical Training School, the Aviation Medicine Unit and RNZAF Base Te Rapa, became responsible to the AOC Support Group. RNZAF Headquarters Singapore and RNZAF Base Shelly Bay came under the direct control of the Chief of the Air Staff, and some specialist units came under technical control of various Defence Headquarters staffs.

A McDonnell Douglas Skyhawk jet fighter.

Air Commodore T. F. Gill headed an aircraft evaluation team which visited the United States in April 1968. When the team left New Zealand, the Phantom, Freedom Fighter and Mirage had already been eliminated on the grounds of cost or operational unsuitability, so the mission's task, in fact, was to negotiate an acceptable Skyhawk purchase deal within prescribed financial limitations. After visiting the Douglas factory in California, where several members flew the A-4, the mission travelled to Washington DC where a satisfactory deal was negotiated.

In 1969 Squadron Leader W. R. Donaldson was sent to the United States on a two-year period of exchange duty to train on the A-4 and then set up a training programme for the RNZAF pilots who would follow him there in 1970. He went to the United States Naval Air Station at Cecil Field, Florida, training with Attack Squadron 44. In December that year he had an unenviable experience which resulted in the loss of an eye. He was flying with a student on a low-level navigation exercise in a TA-4 when a bird, a mallard drake, penetrated the aircraft's canopy and struck Donaldson on the face. Another bird went into the engine. Donaldson heard a loud noise and suffered a sudden loss of vision, but at that stage was unaware of what had happened. The student was flying the Skyhawk at the time and Donaldson spoke to him but, getting no reply from the front seat and being unable to see, he assumed the worst and presumed that he had sole control of the aircraft. They had been flying at 200 feet and at a speed of more than 400 knots. Donaldson initiated a climb and, as the speed reduced, he ejected from the aircraft. He landed in a cornfield, bleeding from the face, quite blind, but otherwise unhurt. He was rescued by a State Highway trooper and taken to Tallahassee Hospital. His pupil was still in the Skyhawk, without electrical instruments or his instructor, so he re-assumed control and made an emergency landing at Tallahassee Airport.

Squadron Leader Donaldson spent three and a half weeks in hospital. One eye was damaged beyond repair and had to be removed but the other recovered sufficiently for him to be able to remain in the Service. The US Navy asked that he be permitted to complete his attachment. Despite his disappointment that his operational flying days were over, Donaldson, with the co-operation of his American colleagues, set up a training programme for RNZAF pilots at Cecil Field.

Ten RNZAF pilots went to Cecil Field on 9 January 1970 for conversion training on the A-4, while 48 ground staff began training there on Skyhawk servicing procedures three weeks earlier. Squadron Leader C. S. H. Calvert was the Training Liaison Officer for the teams at Cecil Field.

The 14 Skyhawks, cocooned for transit, arrived in New Zealand early in 1970 as deck cargo aboard the US aircraft carrier *Okinawa*. Towed along Auckland's north-west motorway to Whenuapai, they were there readied for flight and Squadron Leader J. A. Scrimshaw, Officer Commanding No. 75 Squadron, flew the first aircraft on 21 May. The new Skyhawks were officially handed over at Ohakea on 10 June before the Hon. David Thomson, Minister of Defence, Lieutenant-General Sir Leonard Thornton, Chief of the Defence Staff, and Air Vice-Marshal Stratton, who had succeeded Air Vice-Marshal Turner as Chief of the Air Staff.

The decision to purchase the A-4 met with some criticism within the Service, but it would have been unrealistic to expect total agreement, whatever aircraft was chosen. The Skyhawk, however, met the Air Force's requirements; it had the range to fly to South-East Asia, it had an excellent weapons delivery capability, and it had been well proven in combat in the Vietnam War. Many senior officers who were closely associated with its introduction and its subsequent operation considered the deal with McDonnell Douglas represented the best value for money the RNZAF had ever achieved.

Skyhawks refuelling in flight.

CHAPTER TWENTY-THREE

Service in Vietnam

On 31 March 1962, in a letter addressed to the heads of government of 93 non-Communist states, Ngo Dinh Diem, President of the Republic of Vietnam, drew attention to the serious threat to peace in his own country. The Republic of Vietnam had been compelled to appeal for increased military assistance from Free World countries to prevent its being overwhelmed by massive subversion from Communist North Vietnam, a state backed by heavily increased support from the Communist bloc.

Plagued by political instability, Vietnam had been one of Asia's most troubled countries since the end of the Second World War. The French had fought there from 1946 to 1954 with heavy losses, and partition of the country had followed in July 1954. The Communist forces included the Viet Cong — the indigenous 'national liberation' movement of South Vietnam — and the North Vietnamese Army. Anti-Communist forces in Vietnam included the South Vietnamese Army of the Republic of Vietnam (ARVN).

In May 1961, when Communist insurgents were waging large-scale guerilla warfare in South Vietnam, United States President John F. Kennedy had sent Vice-President Lyndon B. Johnson to Saigon, telling him to proclaim American solidarity with the South Vietnamese. Settling at first on a programme of military aid, the United States progressively increased its support until by 1968–69 there were more than 500,000 American troops in South-East Asia. Between 1964 and 1973 as many as 45,790 were killed, leading to the Vietnam War being labelled as the most unpopular war in United States history.

In November 1963, President Diem was assassinated. The South Vietnamese Government, weakened by recurrent political strife, faced a rapidly deteriorating military situation and the nation appeared to stand on the brink of disaster, with the Viet Cong poised to deliver the death blow.

No RNZAF squadron took part in the Vietnam War as a complete operational entity but 30 pilots served there on individual attachments to USAF or RAAF units, and transport crews of both Nos 40 and 41 Squadrons flew regularly into the war zone.

On 30 January 1967 Cabinet approved the attachment of RNZAF Iroquois pilots to No. 9 (RAAF) Squadron in Vietnam, to serve there for six months. The first officer selected under this scheme, Flight Lieutenant A. R. Mills, went to Vietnam in the second half of 1967, and was succeeded by Flying Officer K. J. Wells early in 1968. Later, authority was given for New Zealand pilots to spend up to two months in orientation training with an RAAF helicopter squadron in Canberra, Australia, before proceeding to South Vietnam. Because of cost, this training phase was subsequently abandoned, but the RNZAF pilots had received first-class training with No. 3 Squadron and were by no means disadvantaged.

On a recommendation from the Air Force, Cabinet, at its meeting on 29 April 1968, authorised both an increased quota of helicopter pilots for Vietnam service and a 12-month rather than a six-month tour of duty.

In the four and a half years from July 1967 to December 1971 a total of 16 RNZAF helicopter pilots served with the RAAF's No. 9 Squadron in Vietnam, operating from Vung Tau in direct support of the 1st Australian Task Force at Nui Dat, some 30 miles south-east of Saigon. The Task Force included two infantry companies, an artillery battery and a troop of Special Air Service Squadron soldiers, all drawn from the New Zealand Army.

Flying Iroquois troop-carrying helicopters and heavily armed Iroquois gunships, the RNZAF pilots took part in combat missions involving the forward supply of troops, medical evacuations, the transfer of ground patrols from forward areas and gunship support of troops fighting on the ground.

Squadron Leader G. Wallingford during service as a forward air controller with the United States Air Force in Vietnam. He was one of 14 RNZAF pilots who served there in this capacity. The aircraft is a turboprop OV-10 Bronco, especially designed for the role.

A further 14 RNZAF pilots served in Vietnam as forward air controllers with the USAF. These men, who spent a short time training in the FAC role at Cam Ranh Bay before taking part in combat operations, directed air strikes and artillery action, marking the enemy's positions. In this work they flew especially designed and equipped counter-insurgency light armed reconnaissance aircraft such as the Cessna 0-2 and the North American Rockwell OV-10 Bronco. The first two pilots to serve as forward air controllers were Squadron Leader J. A. Scrimshaw and Flight Lieutenant R. L. Ewing, who went to Vietnam in 1968.*

High cockpit temperatures on the ground were trying for the pilots and sometimes their sleep was disturbed by rockets hitting the hardstanding close to their quarters, but they found the operational flying generally satisfying and rewarding.

Of the 30 RNZAF pilots who served in South Vietnam, 10 received awards for gallantry.†

Role of the Combined Services Medical Team

In May 1967 a 16-man New Zealand Combined Services Medical Team, drawn from the Army, Air Force and Navy, was sent to Vietnam, to relieve a US Army medical team at Bong Son in Binh Dinh Province, about 280 miles north-east of Saigon. The team's work was directed mainly at the civilian populace, but they also treated military casualties brought to the Bong Son Dispensary, including members of the ARVN and Viet Cong prisoners.

The medical team, which had an administration unit at Qui Nhon, worked in three areas — at Bong Son, 60 miles north of Qui Nhon, where there was an old hospital, soon replaced by a new 100-bed unit; at Phu My, which had a 30-bed dispensary; and at Hoai-An, an isolated village up the Bong Son Valley, where there was an eight-bed dispensary. The New Zealanders did much to upgrade the facilities at Phu My Hospital and they later ceased serving there, working instead at Tam Quan, north of Bong Son. The team comprised three doctors, one administrative officer and 12 medical orderlies, some of whom had specialist qualifications. The hospitals and dispensaries were staffed by Vietnamese personnel. Some could speak a little English but the language barrier did pose some problems.

* Serving in Vietnam as forward air controllers were Squadron Leaders J. A. Scrimshaw and G. Wallingford; Flight Lieutenants M. J. Abel, J. M. Denton, B. W. Donnelly, R. L. Ewing, G. J. W. Goldsmith, R. F. Lawry, B. A. Lockie, D. J. McEvedy, R. J. Metcalfe, G. R. Thomson and P. T. Waller and Flying Officer M. R. Callanan.

† Helicopter pilots who served with No. 9 Squadron, RAAF, included Squadron Leaders G. C. Derby, R. J. Klitscher and J. L. A. Pendreigh; Flight Lieutenants I. McG. Clark, J. B. Clements, E. Y. Creelman, A. R. Mills, G. A. Oldfield, D. I. Paterson, C. R. J. Peters and G. L. Wood; and Flying Officers I. G. Brunton, T. K. Butler, J. D. Peterson, B. H. Senn and K. J. Wells. The DFC was awarded to Squadron Leader Klitscher, Flight Lieutenants Paterson and Donnelly and Flying Officers Butler, Peterson and Wells. The American DFC was awarded to Flying Officer Callanan. Squadron Leader Pendreigh and Flight Lieutenants Metcalfe and Wood were Mentioned in Dispatches.

Squadron Leader R. G. B. Graham, an RNZAF medical officer with the New Zealand Combined Services Medical Team in South Vietnam, examines a young patient in the Phu My dispensary. *New Zealand Army Information Service*

The main purposes of the team's presence in South Vietnam were to assist the Vietnamese in operating their hospitals and dispensaries, to pass on their own knowledge and experience and to help in immunisation and preventive medicine programmes. There were about 42 such teams, military and civilian, in South Vietnam, representing a dozen countries, but the New Zealand team was the only source of external aid available to hospitals in the Binh Dinh province. Because they were helping a seriously underprivileged community, the New Zealanders found the work both satisfying and rewarding and there was a wide range of diseases to test their diagnostic abilities and therapeutic skills. They enjoyed an excellent relationship with the New Zealand civilian surgical team at Qui Nhon, which operated in the war-ravaged country for almost a decade.

Transport Squadrons' Involvement

Throughout the eight years of New Zealand military involvement in Vietnam, the RNZAF transport force maintained regular schedules to supply the various New Zealand units and personnel stationed there. Apart from special flights, No. 41 Squadron — its Bristol Freighters now camouflaged — flew regular weekly schedules for most of the period from 1968 to 1971; the frequency was then reduced after New Zealand infantry withdrew from Vietnam and two of the squadron's Freighters were replaced by Iroquois helicopters. These schedules meant overnighting somewhere in the country, usually at Saigon but often in less pleasant surroundings, and involved operating further up country into marginal airstrips while skirting around Viet Cong-held territory and areas of Allied artillery fire. To the scattered New

A mother and her badly injured daughter at Bong Son Hospital. *Right:* Bong Son dispensary.

Zealanders the weekly Bristol was a welcome sight; to many Americans, a curious one.

No. 40 Squadron also operated to Vietnam throughout the war, with schedules on average of about once a month from New Zealand. These aircraft were used mainly to supply large items of equipment such as replacement vehicles, and to change over large numbers of troops. The uncamouflaged Hercules normally operated only into Saigon or Vung Tau.

The Collapse

When Vietnam fell quickly to the Communists in 1975, this created a delicate situation for the RNZAF. New Zealand maintained an embassy in Saigon and there were other New Zealanders resident in South Vietnam. Obviously the New Zealand nationals would have to be evacuated, and quickly. Just before the collapse Wing Commander J. A. Scrimshaw, who was then commanding the RNZAF in Singapore, was asked to go to Vietnam and assess the situation. First he was required to determine the risks confronting the RNZAF transport crews in their routine flights to Saigon and to other centres such as Qui Nhon. Second, he was to assess how ready New Zealand personnel in Saigon were to evacuate, in the event of a collapse. Wing Commander Scrimshaw said later: 'It was clear at that stage that the Communist forces were approaching Saigon and that they had the ability to shoot down aircraft as and when they desired. It was equally clear that there were no plans for an evacuation of our Embassy and that as it was directly opposite the United States Embassy, evacuation might be difficult.'

On 4 April 1975 an RNZAF detachment of two Bristol Freighters and a servicing party flew from Singapore to Saigon. It was commanded by Squadron Leader R. Davidson, officer commanding No. 41 Squadron. That night the first of six evacuation flights was made to Singapore, with some members of the surgical team from Qui Nhon on board, and the following day one Bristol continued the evacuation while the other began a week-long airlift of rice and bread from Saigon to Phu Quoc, an island in the Gulf of Thailand, where more than 60,000 Vietnamese refugees were gathered. On each of the one and a half hour flights to the island the Freighter had to make a detour around the Mekong Delta to avoid reported anti-aircraft and missile sites.

An RNZAF forward air controller marks target flying a Cessna O-2.

The airlift, which was co-ordinated by the United States Aid Organisation for the Vietnamese Ministry of Social Welfare, also involved the Royal Australian Air Force. One flight was made to Vung Tau with 5000 lb of rice which was unloaded by the crew against a background of small arms fire.

Between 6 and 19 April an RNZAF C-130 detachment under the command of Wing Commander A. E. Thomson, officer commanding No. 40 Squadron, made three trips between Saigon and Singapore, evacuating New Zealand Embassy staff members, refugee children and news media representatives.

Headquarters for the 41 Squadron detachment (known as 'Fort Davidson' or the 'Rice Redoubt') was tented accommodation alongside the control tower on Tan Son Nhut airport. It was fortified with bags of rice, for no one knew when the Viet Cong would overrun the area. Beyond the city five Government divisions faced at least eight Communist divisions, and the morale of the Communists was known to be incredibly high while that of the Government forces was low.

For the first week Fort Davidson was 'home' for up to 16 men — air and ground crews — but with the ending of the airlift of food to the refugees on 13 April and the withdrawal of one Freighter, the remaining crews moved into a large, air-conditioned caravan made available by the Americans, their meals being provided by Flying Tiger Airlines. Working from the New Zealand Embassy in Saigon, Squadron Leader Davidson maintained a close liaison with the ambassador, Mr N. T. Farrell, with Wing Commander Scrimshaw, and with his detachment at the airport, aided by a 24-hour battery-operated portable radio system. This was arranged by Squadron Leader Peter White, a 41 Squadron air electronics officer, with the help of a New Zealand Army radio technician. A radio link back to Singapore was maintained, using the aircraft's own HF radio system.

The evacuation of the New Zealand ambassador and his staff had to be a clandestine operation. If it became known that any one of the three major embassies whose countries had supported South Vietnam in the war — Australia, New Zealand and the United States — was considering total withdrawal, this could have led to widespread public alarm and the situation, already tense, could have deteriorated rapidly. By the weekend of 19–20 April the Communist strength had increased to 13 divisions and Saigon was confronted on all sides except the south-east by North Vietnamese Army divisions and Viet Cong troops, and within the city itself there were South Vietnamese factions whose actions could not be predicted. The final phase of the evacuation was extremely difficult.

On 18 April Wing Commander Scrimshaw issued new orders for the 41 Squadron detachment commander, the key instruction being that the RNZAF party was not to leave Saigon without the embassy staff and other New Zealand nationals aboard. On the morning of Monday 21 April, knowing the Communists were within artillery range of the city, and aware of his responsibility for the RNZAF aircraft and crews, the New Zealand ambassador decided to evacuate all New Zealand nationals who were prepared to leave and as many Vietnamese as possible.

Everyone was tense on the late afternoon flight out of Saigon. The captain of the Freighter, Flight Lieutenant D. Carter, dared not risk supplying the usual manifest as he had Vietnamese passengers aboard, and when the control tower inquired whether clearance formalities had been completed, Squadron Leader White gave the only possible reply and said 'Affirmative'. The hours passed, the crew hot and edgy. Squadron Leader White had a loaded Sterling sub-machine gun beside him and when permission to taxi was given at last, he stood up in the signaller's hatch, watching for likely last-minute hazards. (There were two Vietnamese babies aboard, one of whom, a three-month-old girl, had been adopted by Squadron Leader White.) The existence of ground-to-air, hand-held, heat-seeking missiles made the crew uneasy, but the Freighter took off without incident and completed an uneventful flight to Singapore.

Thus, on 21 April 1975, the RNZAF closed another eventful chapter of its history.

The BAC 167 Strikemaster advanced trainer introduced into service in 1972.

CHAPTER TWENTY-FOUR

Hail and Farewell

There were many significant changes in the Royal New Zealand Air Force in the 1970s, a decade notable for the reappearance at Auckland of the RNZAF's oldest squadron — No. 1; the re-equipment of Nos 14 and 42 Squadrons at Ohakea; the introduction of New Zealand-built basic trainers at Wigram; and the disbandment of No. 41 Squadron after 33 years of outstanding service. These years also saw the integration of airwomen into the RNZAF.

Reactivation of No. 1 Squadron

At RNZAF Base Auckland on 1 August 1972, No. 1 Squadron was reformed under the command of Squadron Leader P. R. Adamson. The unit, equipped with Bristol Freighters, had previously been the medium-range/tactical transport flight of No. 3 Squadron. The latter unit could now concentrate on helicopter operations, while No. 1 Squadron continued with the tasks of supply dropping, internal transport, overseas transport and the training of crews for No. 41 Squadron in Singapore.

Re-equipment of No. 14 Squadron

No. 14 Squadron began to re-equip in 1972 when the purchase of 10 BAC 167 Strikemaster aircraft provided a new advanced trainer and enabled the Vampires to be phased out. A further six Strikemasters were later purchased and in 1976 the squadron began its new role of advanced flying training for student pilots and operational training for pilots earmarked for strike duty. The squadron's forward air control Harvards were retired in 1977. The Strikemasters were new aircraft when purchased and in their first 13 years of service flew almost 45,000 hours without an accident. (The first Strikemaster to be lost in a flying accident crashed in 1985, but without loss of life.)

Andover Purchase

In 1976 the Chief of the Air Staff, Air Vice-Marshal R. B. Bolt, learned from his RAF counterpart, Air Chief Marshal Sir Andrew Humphrey, who was visiting New Zealand, that surplus Hawker Siddeley Andover aircraft were available in the United Kingdom. As replacement aircraft were urgently needed for the aging Bristol Freighters and the much older Dakotas, he put a proposal to the New Zealand Government to buy the Andovers. The British CAS assisted in the negotiations for an acceptable package and in July that year the Government announced the purchase of 10 Andovers at a cost of $13 million.

The twin turboprop Andover C.1 was a military derivative of the HS748 airliner, and had been built during 1965-67 to meet the needs of the Royal Air Force for a STOL (short take-off and landing) tactical transport aircraft. Four of the 10 were allocated to 42 Squadron, then commanded by Squadron Leader R. L. Randal, two being refurbished by No. 1 Repair Depot at Woodbourne to the standards required for the carriage of VIPs. The other six were allocated to No. 1 Squadron and its Bristol Freighter aircraft were withdrawn from service in 1977. The utility version Andovers which replaced them had a cruising speed of 210 knots and a range of 1100 nautical miles with maximum fuel and 8500 lb of payload. They could carry 44 passengers with light baggage. As well as assuming the routine transport tasks previously flown by Bristol Freighters, No. 1 Squadron's Andovers were also for a time employed on regular patrols to police New Zealand's 200 mile economic zone.

The Andover purchase not only facilitated the phasing out of the Dakotas and Bristol Freighters but also helped to restore morale to No. 42 Squadron. Throughout the late 1960s and early 1970s the air transport of VIPs — traditionally the role of 42 Squadron — had been entrusted more and more to civilian operators. The VIP Dakotas, once the pride of 42 Squadron's fleet, were obsolete, slow and noisy, while NAC's Friendships and those of Civil Aviation Division's Calibration Flight were

Air Marshal Sir Richard Bolt, KBE, CB, DFC, AFC, Chief of the Air Staff 1974 to 1976, Chief of the Defence Staff 1976 to 1981. He was the first New Zealander to achieve the rank of Air Marshal while serving in the RNZAF.

An Andover of No. 1 Squadron, still in RAF camouflage, shows off its rough airfield landing capability. The aircraft has just touched down; the camera has 'stopped' the propellers.

A forward air control Harvard of No. 14 Squadron over the Crater Lake, Mount Ruapehu, in the 1970s.

The last Devon in flying service with the RNZAF, NZ1822, (below left) was withdrawn in August 1981 following the introduction of the Cessna Golden Eagle, represented by NZ7941 (centre). To mark the occasion No. 42 Squadron flew this formation with Andover NZ7627 leading it.

newer, faster and quieter. Inevitably, while VIPs were carried in civil aircraft, morale within 42 Squadron suffered. The squadron's unit history for May 1976 recorded that no VIPs had been carried in the previous seven weeks!

After 42 Squadron's first Andover went into VIP service it quickly proved its worth. Talks between the Department of Internal Affairs and the RNZAF during 1978 confirmed the belief that Government-hosted tours would make ever increasing demands on the Air Force, and that there would be plenty of work for the second VIP Andover when it came into service in the middle of the year. This expectation was fully borne out. Mr Walter Mondale, Vice-President of the United States, was one of the first foreign visitors to be carried and from then on the squadron flew more and more VIPs.

When the first Andover, NZ7622, arrived at Ohakea on 5 February 1977, it was escorted in by two Dakotas. Throughout 1977 the long-serving Dakotas were retired progressively until, on 12 November that year, Dakotas NZ3551 and NZ3553 flew past ceremonially at Ohakea before the Chief of the Air Staff, Air Vice-Marshal C. L. Siegert (who in 1953 had flown the Queen in NZ3551), wartime Wing Commander F. J. Lucas, who had pioneered the transport routes in the first RNZAF Dakota 34 years before, and 400 other guests. (With Government approval, the 'Royal' Dakota NZ3551 was allotted to the RNZAF Museum for preservation as an historic aircraft.)

The tasks for 42 Squadron grew in number month by month. February 1979 saw a record 139 hours flown on VIP transportation and the Maintenance Flight worked long hours so that the demanding schedule could be met.

Re-equipment at Wigram

When the Flying Training Wing at Wigram began re-equipping in 1976 so that the long-serving Harvards and Devons used for pilot training could be phased out, it was a New Zealand-built aircraft that was chosen — the CT-4B Airtrainer, manufactured by New Zealand Aerospace Industries at Hamilton. Thirteen were delivered that year and another six in 1978.

There was no controversy about the choice. The Air Force was already committed to the purchase of the

The VIP Andover of No. 42 Squadron.

Strikemasters and needed a basic trainer, particularly one with an instrument flying capability, that would take pilots halfway through their training. Both Air Vice-Marshal Bolt and his predecessor Air Vice-Marshal D. F. St George were keen to consider the Airtrainer, thus supporting the development of New Zealand's own aviation industry.

In July 1976 the first RNZAF Airtrainer was test flown and delivered to Wigram by Flight Lieutenant N. A. Davey, the RNZAF's acceptance pilot for the programme. Pilots at CFS were quickly converted on to the aircraft and the staff began writing a new flight manual and a flying instructors' handbook for the type. Because of the usual teething troubles with a new aircraft and a lack of some technical items, the Air Staff decided in November 1976 to delay the introduction of the Airtrainer until June 1977. The six months' breathing space allowed many of the problems to be overcome and the others were to be removed by the ingenuity of air force technicians in the years ahead.

The veteran Harvards, which had been in RNZAF service for 36 years, could now be honourably retired. A flight of 17, led by Wing Commander D. G. McAllister, MBE, officer commanding the Flying Training Wing, gave a farewell flypast on 24 June 1977.

While the Airtrainer went into service for pilot basic training and the training of flying instructors, navigators and air electronic operators continued training on the remaining Devon aircraft until three Fokker F-27 Friend-

Strikemasters of No. 14 Squadron at Ohakea.

Members of the Red Checkers aerobatic team put their Airtrainers through their 'mirror image' manoeuvre.

The New Zealand-built CT-4B Airtrainer, introduced in 1977 as a basic trainer.

ships were acquired during 1980. The last Devon navigation training flight was made on 18 April that year, bringing to a close almost 28 years' service in the role.

Because the newly acquired Friendships had been in service with the national airline Air New Zealand (and the former National Airways Corporation) for 19 years, some thought their purchase a retrograde step. Modifications, however, included the fitting of a later model instrument panel, Doppler and Omega navigation systems, underwing pylon fuel tanks, new avionics, a periscopic sextant, a navigation station for two trainee navigators and a station for a navigator instructor and air electronics operator. With a completely refurbished interior and the ability still to carry 24 passengers, the Friendship provided the RNZAF with a multi-role aircraft. Despite early problems with the calibration of the new compass and Omega systems, the three Friendships proved reliable and have continued to be popular with crews.

Equality for Airwomen

The change of policy by which airwomen were integrated into the RNZAF to serve on an equal basis with men became effective on 29 July 1977, and although the Women's Royal New Zealand Air Force was not immediately disbanded, being retained as 'an inactive entity', it virtually ceased to exist for women then serving. All Air Board orders and other standing instructions were amended to remove discriminatory clauses, and the new policy specified that women should be excluded only from those trades which primarily required the use of weapons for effective performance or involved a significant amount of physically demanding work. These exclusions meant that women could not be employed as aircrew.

Because the new policy provided for the exposure of servicewomen to defensive combat in an emergency, subsequent recruit and NCO courses included basic training for women as for men in the use of small arms.

The Passing of 41 Squadron

After almost 26 years of continuous operations, No. 41 Squadron's Bristol Freighters were more than ready for retirement, but No. 41 Squadron, which had flown the aircraft since 1951, was not to survive either, and 1977 was to see this unit removed from the RNZAF's Order of Battle.

The Fokker F27 Friendship.

On 1 November 1971 the Anglo-Malaysian Defence Agreement was terminated and was replaced by the Five Power Defence Arrangements, in which New Zealand, Australia and the United Kingdom separately concluded agreements with Malaysia and Singapore. Within the framework of these agreements and at the request of the two South-East Asian countries, Commonwealth forces were to continue to be stationed there. It was agreed between Australia, New Zealand and the United Kingdom that, for administrative and logistic convenience, their forces would be grouped into an 'ANZUK' force.

No. 41 Squadron accordingly became a part of New Zealand's contribution to the ANZUK Force; earlier it had moved its base from Changi to Tengah, which in September 1971 became a Republic of Singapore Air Force base. The ANZUK Force's role was to contribute to stability while both Singapore and Malaysia built up their own defence capabilities, and it was implicit in the new arrangement that New Zealand would maintain forces in the region only while their presence was welcome and considered valuable to all parties.

At Tengah the Freighter component of No. 41 Squadron was reduced to two aircraft while helicopters were added to the inventory for the first time. When three Iroquois joined the unit, this brought about a vast change to the nature of operations. Late in 1972 the squadron strength was increased by a third Bristol Freighter and a fourth Iroquois.

During 1974 the ANZUK Force was run down and its headquarters was disbanded in January the following year. In December 1974 No. 41 Squadron reverted to national command. Inevitably, there was some change in the work of the squadron. January 1975 marked the first of a regular schedule of flights to East Malaysia in support of the Malaysian Army, and in April that year 41 Squadron's involvement in Vietnam came to an end with the collapse of the Saigon Government.

Despite the New Zealand Labour Government's announcement that the New Zealand forces would be withdrawn within two years the squadron remained in the region and continued to diversify its tasks.

In 1977 the Defence Council decreed that the Bristol Freighters should be retired and No. 41 Squadron disbanded. On 6 December, after a number of formal and informal events to mark the occasion, No. 41 Squadron paraded for the last time, under Wing Commander G. A. Parkinson. Elements of the New Zealand Force in South-East Asia paid their respects by associating two companies of the First Royal New Zealand Infantry Regiment and the Band of 1 RNZIR with the parade. Aircraft of the Royal Australian Air Force and the Republic of Singapore Air Force flew past. After 33 years of varied and distinguished service in many parts of the world, No. 41 Squadron was no more.

On 15 December the 41 Squadron Standard was laid up in the chapel at RNZAF Base Auckland after being marched through Auckland streets lined by Air Force men and women from the base, with former members of the squadron looking on.

The decision to disband 41 Squadron was taken partly because of the need to retire the Bristols but also because of the diminishing requirement for a fixed-wing transport element based in Singapore. A contributing factor, too, was the existence of No. 1 Squadron at Auckland in the tactical transport role. Future 'in-theatre' fixed-wing transport tasks would be met by Hercules and Andover detachments from New Zealand, operating extended schedules. The new unit built around the Iroquois helicopters remaining at Tengah with 79 RNZAF personnel and 38 locally employed civilians was given the title RNZAF Support Unit, Singapore (RNZAF SUS).

Countless New Zealanders have had cause to be grateful for the search and rescue operations of No. 3 Squadron's Iroquois helicopter crews.

An Orion of No. 5 Squadron on a Pacific search and rescue mission.

CHAPTER TWENTY-FIVE

The Global Odysseys of No. 40 Squadron

During the 1970s No. 40 Squadron could lay an even greater claim to a worldwide role as the five Hercules were flown to places the RNZAF had never before visited. As a December 1973 entry in the squadron history states:

> While Wing Commander M. W. Hodge and his crew were flying within Soviet airspace at approximately 57 degrees north, Squadron Leader I. W. Collins and his crew were on the ground at Williams Field, Antarctica, 78 degrees south. South America presently remains the only continent that has not been visited by 40 Squadron's aircraft. Once this milestone has been reached,* the squadron motto, "Ki Nga Hau E Wha" ("To the Four Winds") will have been realised.

In December that year alone the squadron flew nine return flights to Antarctica, made an historic flight into the USSR, flew relief aid into Fiji following Cyclone Lottie, assisted No. 5 Squadron's deployment from RAAF Edinburgh after the Fincastle Trophy competition, maintained the usual internal schedules and carried the Prime Minister of New Zealand, the Rt Hon. Norman Kirk, on a tour of South-East Asia.

From its reformation with Handley Page Hastings in 1954, No. 40 Squadron began flying regularly to Fiji, Singapore, Great Britain and the Middle East. The addition of DC6s in 1961 saw the North American continent added to the squadron's regular routes. In the 1970s new tasks found the Hercules crews flying into many other countries for the first time, notably China, the Soviet Union, Bangladesh, Rhodesia (now Zimbabwe), Kampuchea and Iran. When three Boeing 727 jet transports† were added to the fleet in 1981, the squadron's global operations were further expanded.

Aid to Bangladesh

When the new independent nation of the People's Republic of Bangladesh was formed in 1972, the 54,000 square miles of territory that had once been East Pakistan lay in a state of chaos. The retreating military forces of West Pakistan left a trail of destruction across an already poverty-stricken country. Entire towns were destroyed, leaving millions of people homeless. To add to this confusion, refugees were wandering the country in aimless dejection.

* Such a flight was being planned in 1986.

† These aircraft ex-United Airlines were purchased in the USA.

The first priority of the Government of Bangladesh was to keep its 75,000,000 people alive. This task required food in huge quantities and a means of transporting it through a country whose normal communications network was in ruins. Because this massive task was clearly beyond the resources of Bangladesh itself, it looked for outside assistance.

New Zealand was one of the first nations to formally recognise the new state of Bangladesh, and the Government acted quickly to contribute to an international emergency aid programme. No. 40 Squadron was tasked for the mission, giving the RNZAF its first involvement in a major international relief operation within a foreign country.

Two separate detachments were sent. Squadron Leader N. J. S. Rodger commanded the first party, leaving Whenuapai on 23 February and returning on 13 March. Squadron Leader Rodger's aircraft carried more than 1,000,000 lb of urgently needed food throughout Bangladesh in a 14-day airlift, and the success of this mission led to the dispatch on 26 March of a second aircraft commanded by Squadron Leader P. W. Tremayne, which spent three weeks airlifting supplies which included rice, wheat, sugar and milk powder. In 73 separate sorties the second Hercules carried 3,135,604 lb of relief foodstuffs before returning to New Zealand on 20 April.

In essence, the New Zealanders' task was the movement of vital foodstuffs from the sea ports and railheads into the interior of Bangladesh. The aircraft flew several routes — from the principal port of Chittagong to the airfield at Dacca, from Dacca to the northern rail terminal of Isheerdi (in the north-west sector of Bangladesh) and from Baghdogra (an Indian Air Force airfield in Northern India) to Isheerdi. The normal offload airfields were Dacca and Isheerdi.

The crew of the day would begin duty at 4 o'clock in the morning and finish at 8.30 at night. Because of airfield damage, they could not fly at night and had to take full advantage of the hours of daylight.

The work was demanding and arduous. Each New Zealand crew would try to stage six return flights a day, flying in and out of war-damaged airfields and sweating in temperatures of up to 43°C while loading and off-loading huge quantities of supplies. This type of operation continued for 20 days without a break. Flying conditions were often extremely hazardous. A feature of the local weather was late afternoon thunderstorms of great severity, which developed rapidly, reaching altitudes of

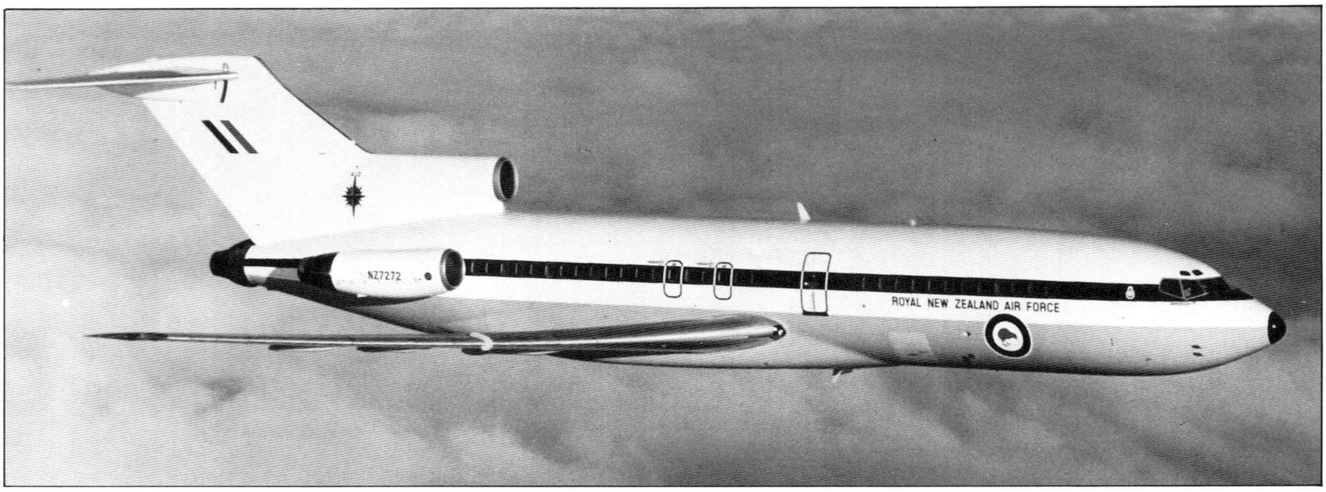

A Boeing 727 of No. 40 Squadron in flight.

60,000 feet or more. They resulted in airfield flooding, failure of electric power, flying debris and general disruption of radio navigation equipment.

This was not a military aid programme. The RNZAF Hercules were, in fact, the only military aircraft involved in the emergency operation and all personnel were issued with Red Cross identification cards. Because the seven civil aircraft flying under charter to the International Red Cross carried smaller loads (8 tons per flight compared with the Hercules' 20 tons), the two C-130s made a significant contribution, which brought high praise from many quarters and earned the RNZAF the Red Cross Medallion for Meritorious Service.

Mission to China

In July 1973 Wing Commander M. W. Hodge captained Hercules NZ7002 on an historic flight into the People's Republic of China. The squadron had been given the task of transporting furniture and equipment into Peking, now known as Beijing, for the establishment of a New Zealand embassy there. Leaving Whenuapai on 22 July, the aircraft staged through Townsville and Guam to Hong Kong where final preparations were made for the flight across the Chinese border. The aircraft stopped briefly at Canton to complete formalities for entering Peking and to uplift a Chinese navigator, radio operator and interpreter to assist the crew on the final leg of the journey. The Chinese navigator worked alongside Squadron Leader N. J. Barr, navigator for the outward flight, while the Chinese radio operator assisted Flight Lieutenant C. W. Adamson, the co-pilot, with ground communications that were carried out in both English and Chinese. Throughout China the aircraft was under the control of the Civil Aviation Administration of China.

The flights across China gave the crew their first experience of operating in a totally metric airspace. As the aircraft crossed the border from Hong Kong its altitude was converted from feet to metres and its airspeed from knots to kilometres an hour. Heavy rain greeted the arrival of the Hercules at Peking on 25 July, after 25 hours in the air since leaving New Zealand. Next morning the aircraft began the return flight, and by the time it had reached Whenuapai it had travelled 15,000 miles in a little over 60 hours' flying time.

Into the Soviet Union

Twilight was fading and snow falling heavily as Hercules NZ7002 landed at Moscow's Sheremetievo airport at the end of a 17-hour flight from England. It was just after 4 p.m. on 5 December 1973 and a New Zealand military aircraft had landed in the Soviet Union for the first time.

An instrument approach had been necessary because of bad weather in and around Moscow and as Wing Commander M. W. Hodge had broken cloud he and his crew had been confronted by a fairytale scene of Moscow in the failing light under a mantle of snow. The temperature outside was -18°C.

The aircraft was met by New Zealand embassy officials and two Russian security men. The embassy staff had a special interest in the flight as the Hercules carried furniture and equipment for the setting up of a New Zealand embassy in the Russian capital; this was the first of three such flights for the squadron between Lyneham in the United Kingdom and Moscow. The New Zealanders flew via Copenhagen in Denmark where the aircraft was refuelled and a Russian navigator and radio operator, both

Wing Commander M. W. Hodge (left) and Flight Lieutenant T. K. D. Reynish with their two Russian crew members at Kastrup Airport, Copenhagen, in 1973.

A rare moment in the life of No. 40 Squadron — five Hercules transport aircraft and one Boeing 727 flying in formation.

seconded from Aeroflot, were taken on board for the flight into Moscow through Soviet airspace.

Cyclone Tracy

In December 1974 No. 40 Squadron was alerted for a relief mission to Darwin on the northern coast of Australia where a cyclone of devastating ferocity had left thousands of residents homeless. Cyclone Tracy levelled Darwin on Christmas morning and 48 hours later, when the Australian Government had a massive relief operation mounted, the known death toll was 49 and survivors were recovering from their second night of sleeping in emergency accommodation.

The suddenness of the disaster and the extent of the damage had a stunning effect on the peoples of both Australia and New Zealand. While Australian Prime Minister, Mr Gough Whitlam, was hurrying home from a visit to Great Britain, his deputy, Dr Jim Cairns, was arranging an emergency airlift of survivors. Seasonal festivities were suspended in many areas as 'Darwin Priority' signals alerted the people who would be needed for the rescue operation.

New Zealand's Prime Minister, the Rt Hon. Wallace Rowling, did not wait for a request for help but offered air transport at once, a gesture the Australian Government was quick to accept. Wing Commander A. E. Thomson, who had assumed command of No. 40 Squadron at the beginning of 1974, lost no time in readying an aircraft and selecting two crews. By mid morning on 27 December the C-130 was on its way to RAAF Base Richmond, headquarters for the relief operation. Over the next five and a half days the New Zealand crews flew for more than 65 hours, carrying urgently needed equipment into Darwin and evacuating people from the town. This was taxing work for the RNZAF crews. Like most emergency missions the airlift called for concentrated, round-the-clock flying. Inevitably, crew members suffered from fatigue.

Most of the residents of Darwin had to be evacuated because the vital services of the town were out of order. Communication was re-established by the use of an Air Transportable Telecommunications Unit which was flown into Darwin as soon as the town's runway could be used with safety. For the Royal Australian Air Force, the Cyclone Tracy emergency operation was its greatest contribution to civilian relief following a natural disaster. By 31 December a total of 35,363 people had been evacuated from the Northern Territory capital, 25,628 of them flown in Service and civil aircraft.

The RNZAF detachment was commanded by Flight Lieutenant G. J. Fallow, who was also a crew captain; the other crew was captained by Flight Lieutenant G. J.

Wood. Most of the residents flown out of Darwin by the RNZAF detachment were women and children, all of whom carried pathetically few personal belongings. Because of the preponderance of children it was not uncommon for the 84-seat Hercules to take off with 120 people on board. The RNZAF Hercules was released by the Australian National Disaster Board on 2 January 1975.

Flights into Burma

In 1977 and 1978 No. 40 Squadron flew into Burma with machinery and engineering supplies for a New Zealand-sponsored technical education project. Squadron Leader P. Bevin captained the first Hercules, leaving Whenuapai on 27 January 1977 with the officer commanding No. 40 Squadron, Wing Commander M. E. A. Dillon, on board as mission leader, flying into Myitkina with an ambulance and machinery for the school there. Two further flights to Myitkina were made in March 1978, the Hercules on those occasions being captained by Squadron Leader C. W. Adamson. On board for the first of the March flights was Mr Alex Owen, head of the technical department at Taita College and technical adviser to the Ministry of Foreign Affairs, who spent nearly eight years in Burma helping to establish technical education there. At the end of each airlift the Hercules returned to Rangoon for an overnight stop before flying on to Singapore.

Squadron Leader I. G. Brausch made another flight to Burma on a similar mission two years later.

'Operation Icecube'

After the first three Hercules went into service in 1965 No. 40 Squadron flew regularly into Antarctica, touching down on the ice at Williams Field, to fly in personnel and materials for New Zealand and United States research programmes there. Air Force involvement in Antarctic research began in 1956 with the formation of the Antarctic Flight* and between 1961 and 1964 several RNZAF pilots were attached to a US Deep Freeze support squadron, operating first with Dakotas and Otters and later with ski-equipped C-130s. These men were later posted to No. 40 Squadron and when 'Operation Icecube', as RNZAF participation in the Deep Freeze programme was nicknamed, began, their experience was invaluable.

Usually, 10 to 12 Icecube missions have been flown each year. As Antarctica is a remote continent with a harsh environment and comparatively few facilities, these operations have been challenging and demanding for the squadron. All crew members — pilots, navigators, engineers and loadmasters — have received field training in Antarctic survival at a location near Scott Base, on a two and a half day course, under the direction of trained mountaineers employed by the DSIR for the summer season.

New Zealand's Vanda Station, a cluster of huts in the geographic centre of the Dry Valleys, was dropped into the Wright Valley by No. 40 Squadron in October 1967. Wing Commander B. J. O'Connor, who was then commanding the squadron, delivered the 43,000 lb load of prefabricated sections on combat-expendable-platforms in three separate airdrops on 16 and 17 October. Replacement buildings were dropped there by the squadron in November 1984 when Squadron Leader M. W. Sinclair, AFC, captained a Hercules on a flight to Antarctica carrying 16,000 lb of prefabs.

* See Chapter 20.

A Hercules aircraft of No. 40 Squadron at Myitkyina, Burma, in 1978, after airlifting machinery and tools to the Burmese city. Burmese soldiers show a lively interest in the aircraft.

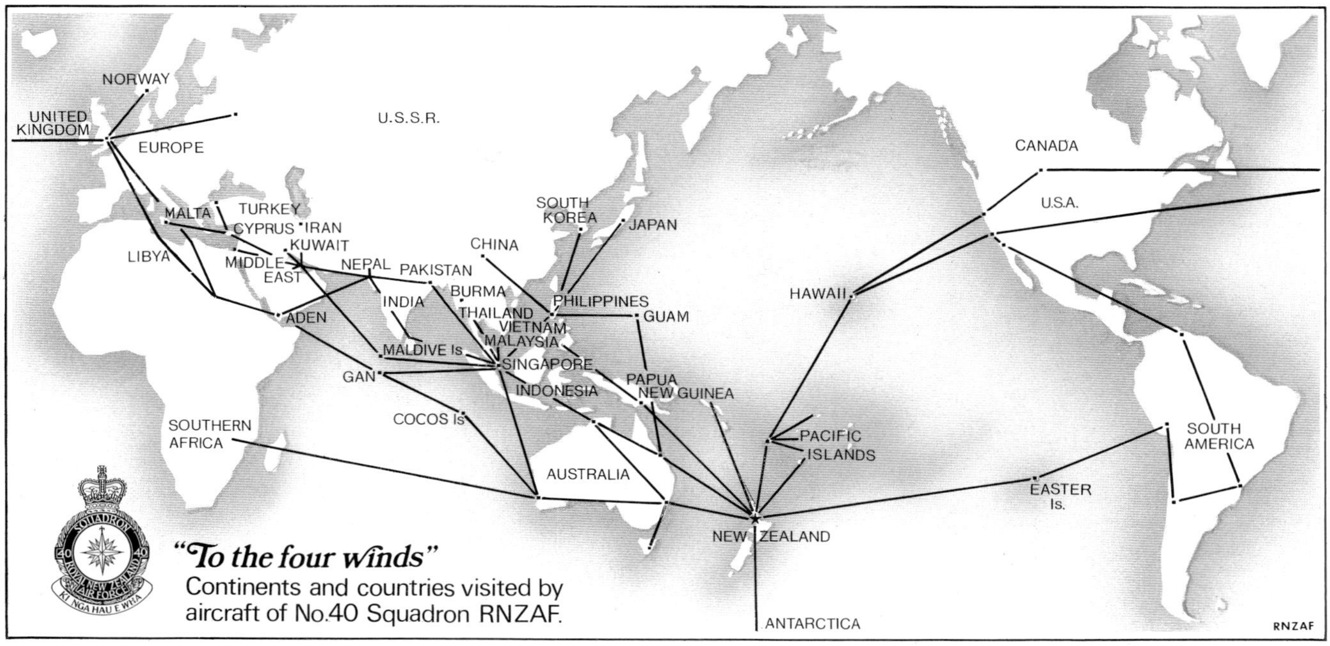

"To the four winds"
Continents and countries visited by aircraft of No.40 Squadron RNZAF.

Other Missions

No. 40 Squadron evacuated New Zealand nationals from Teheran as the Iranian revolution gathered momentum in 1979 and in the same year a Hercules captained by Flight Lieutenant D. S. Glendinning airlifted a truce monitoring contingent of the New Zealand Army into Zimbabwe.

In 1979 40 Squadron gained international recognition by winning the Concours d'Elégance Trophy for the best presented Hercules aircraft at an international air tattoo at Greenham Common, England, celebrating the 25th anniversary of the first flight of the C-130.* The line up of 26 Hercules aircraft represented 14 nations. Hercules NZ7001, 14 years (and 10,000 flying hours) old, commanded by Squadron Leader T. K. Butler, DFC, was judged the best presented C-130 on display. Warrant Officer S. B. Peyton, in charge of the aircraft's maintenance, won the trophy for the best crew chief.

In late 1969 No. 40 Squadron played a major role in an emergency airlift across Cook Strait. Codenamed Pluto 1, the operation was mounted at the request of the New Zealand Government when a seamen's strike left travellers and vehicles immobilised at Wellington and Picton. A similar situation in November 1971 saw two Hercules of 40 Squadron and two Bristol Freighters of No. 3 Squadron, in Operation Pluto 2, carry 1330 passengers, 631 motor vehicles and 35 motor cycles between Woodbourne and Wellington in 280 flights. The efficiency of these operations led to much favourable comment in newspaper editorials, and in subsequent years other such emergencies have resulted in similar operations.

On 29 November 1979, after news was received that a DC-10 airliner of Air New Zealand had crashed on Mount Erebus in Antarctica with 257 people on board, a Hercules of No. 40 Squadron captained by Flight Lieutenant Glendinning flew to Christchurch, took on board a police contingent and an air accidents investigation team, and flew to McMurdo via the crash site. This was a harrowing mission for all on board. When the recovery party had disembarked Flight Lieutenant Glendinning and his crew were required to fly back to Christchurch as there was no accommodation for them at McMurdo. Fatigue symptoms were pronounced in the late stages of their 28 hours' flying operation; normally two crews would have been assigned to such a mission.

A very different mission and one which further illustrates both the versatility of the C-130 and the range of 40 Squadron's activities was the airdrop of a bulldozer onto lonely Pitcairn Island in May 1983. Hercules NZ7005, captained by Squadron Leader T. K. Butler, DFC, flew to Tahiti on 30 May, then flew to Pitcairn the next day and airdropped the 23,000 lb bulldozer in two separate loads. Both drops were made onto a small plateau above the island's settlement. The mission was flown in response to a request from the British High Commission.

During the 1970s 40 Squadron flew numerous missions to Pacific Island countries, for cyclone relief and for emergency medical evacuations. Several of these missions were flown in association with No. 3 Squadron, whose Iroquois helicopters could be carried aboard C-130 aircraft, together with 3 Squadron maintenance staff and aircrews.

When Hurricane Meli swept through the Fiji Group late in March 1979 both squadrons were alerted for a relief mission and a No. 3 Squadron detachment, under Wing Commander R. J. Klitscher, was fully occupied in the disaster area for an exhausting two weeks. There were similar calls on the two squadrons — and on Andovers of 1 Squadron and Orions of 5 Squadron — during the 1980s, and with each mission the Fijians' respect for the RNZAF increased. Despite the many emergency calls, No. 40 Squadron continued its routine transport tasks year after year, moving passengers, equipment and freight throughout New Zealand — and to the four winds.

* The prototype YC-130 Hercules aircraft made its maiden flight from Burbank, California, in August 1954.

Marshal of the Royal Air Force the Lord Elworthy, KG, GCB, CBE, DSO, MVO, DFC, AFC, RAF (Retd), New Zealand's highest-ranking airman, who was Chief of the Defence Staff in Great Britain from 1967 to 1971, was elected Patron of the Air Cadet League in 1980. He is shown presenting the ATC with a banner from the league to mark its 40th anniversary. The colour bearer is Pilot Officer Philip Hayward.

Mr H. L. Thomas, an original member of No. 485 (New Zealand) Squadron, who succeeded Air Vice-Marshal Sir Arthur Nevill as President of the Air Cadet League of New Zealand in 1976. In 1981 he was awarded the Queen's Service Order for his services to youth.

ATC cadets of No. 47 (District of Hawera) Squadron on a conducted tour of Waiouru Military Camp and the Army Museum in March 1983. *Hawera Star.*

CHAPTER TWENTY-SIX

A New Shape for the ATC

The 1970s also brought great changes within the Air Training Corps movement. When the ATC was formed in 1941 it was seen primarily as a 'feeder service' for the RNZAF, a kind of junior air force that would ensure a supply of partially trained new recruits once the boys came of age. An all-male organisation, it was financed wholly from the Vote-Air within the Defence allocation by Government.

Early in 1971 the New Zealand Government proposed that the cadet forces — the Army Cadets, Sea Cadets and Air Training Corps — be disbanded and discontinued. (This was an economy measure designed to leave more finance available for military hardware. The proportion of the Defence Vote earmarked for the cadet forces at that time was approximately $400,000.)

When the Government's intention became known, several ex-Service organisations,* led by Air Vice-Marshal Sir Arthur Nevill, then National President of the New Zealand Air Force Association, formed a deputation to the Government and their submissions over the next few months were successful in reversing the decision. The Cadet Forces would remain, no longer as an integral part of the Services but as youth movements supported by the communities in which they were located or by the community at large. The Air Training Corps needed a civilian organisation to shoulder some of the burden previously borne by the Ministry of Defence — an equivalent of the existing Navy League.

Thus came into being the Air Cadet League of New Zealand. Group Captain C. C. N. Johnston, MBE, RNZAF (Rtd), the Dominion Secretary of the Air League, worked with Sir Arthur Nevill to form the organisation and when, late in 1971, an interim national council had been set up, Squadron Leader G. Bentley, MBE, RNZAF (Rtd) was appointed general secretary of the Air Cadet League.

The first years were difficult as most of the local committees acting in support of ATC squadrons did not want to be part of a new national organisation which they saw as a body superimposed from above. As the concept of essential civilian participation became better understood the league gradually became accepted and the ATC flourished within the limitations of available finance. Flying scholarships were introduced for cadets — in addition to the existing RNZAF scholarships — gliding

*The New Zealand Air Force Association, the New Zealand Returned Services Association, the New Zealand Federation of Brevet Clubs and the Air League of New Zealand.

Air Training Corps gliding camp.

tuition was expanded and new squadrons began to appear. By now girls had been admitted to membership on the same basis as male cadets, and they performed well in all classes of training.

Mr H. L. Thomas, one of 485 Squadron's foundation members in the Second World War, succeeded Sir Arthur Nevill as Dominion President. By 1976 there were new ATC squadrons in Balclutha, Whakatane and Kaikohe and in the 1980s further squadrons were established at Auckland, Morrinsville, Rotorua, Porirua, Christchurch and Alexandra. In 1980 New Zealand-born Marshal of the RAF the Lord Elworthy* accepted the office of Patron of the League. He took a keen interest in the cadets' activities, attending various events including the annual flying camps, usually held at Omaka, near Blenheim.

The 40th anniversary of the formation of the ATC was marked in 1981 by Lord Elworthy presenting a banner to the corps on behalf of the Air Cadet League. Members of No. 1 (Wellington) Squadron and No. 2 (Petone) Squadron, the two oldest units, formed the nucleus of a parade at Trentham. Air Vice-Marshal D. E. Jamieson, Chief of the Air Staff, represented the RNZAF.

Over the years a succession of retired Air Force officers guided the destinies of the movement, working in conjunction with the incumbent Commandant of the New Zealand Cadet Forces. By 1986 the ATC had a total strength of 3000 cadets, male and female, training in 50 town units from Kaitaia to Invercargill. Despite a shortage of cadet officers in some towns, the ATC was generally in good heart, with waiting lists in some squadrons.

Although the ATC is no longer constitutionally a part of the RNZAF, the two remain spiritually inseparable and it has been Air Staff practice in recent years to extend to the movement every possible support.

* Marshal of the Royal Air Force the Lord Elworthy, KG, GCB, CBE, DSO, MVO, DFC, AFC, of South Canterbury, was Chief of the Air Staff, RAF, from 1963 to 1967 and Chief of the Defence Staff in the United Kingdom from 1967 to 1971.

CHAPTER TWENTY-SEVEN

Updating the Orions and Skyhawks

New Zealand's geographical isolation in a vast expanse of ocean has always meant that the RNZAF has had to give as much priority to specialised maritime defence as to its strike and transport capabilities. The oceanic region of immediate concern is vast, extending from the Equator to Antarctica — an area equivalent to the land masses of Europe and Africa combined.

The South Pacific can appear a comparatively tranquil place when viewed against the world's trouble spots, but in the late 1970s it was a region of quickening political and socio-economic change, which included a number of emerging mini-states, all with their own problems. In the early 1980s there was evidence of increasing outside interest and more externally directed activity in the area.

These factors made it imperative that the RNZAF retained the ability to detect, identify and report surface and sub-surface activities throughout the entire area of responsibility and to apply force where necessary to deter any challenge to New Zealand's national interests. Protection and control of the fishing and mineral resources within New Zealand's exclusive economic zone also constituted an important part of the Air Force's maritime role.

In the first 12 years of their operation in this role the Lockheed Orion aircraft of No. 5 Squadron gave excellent service, but by 1978 it was acknowledged that, although the basic airframe and engine combination was sound and had an effective operational life extending until the end of this century, the surveillance equipment was obsolescent and difficult to maintain. The RNZAF either had to purchase new aircraft or upgrade the Orion's mission equipment.

When replacement was found to be too expensive, a two-phase modernisation programme was decided on. Phase One comprised an update of the aircraft's navigation and surveillance equipment and the installation of a data handling and display system designed to allow for easy integration of the proposed Phase Two equipment. The second phase would upgrade the Orions' anti-submarine warfare and communications systems by installing replacement acoustics and electronic warfare sensors, providing a stand-off weapon capability, and modifying the communications system to create a data link capability.

In July 1980 the Boeing Aerospace Company was awarded a contract for Phase One of the modernisation programme. Wing Commander R. F. Parker went to Seattle that year, accompanied by Wing Commander P. A. Sherson, to head a special RNZAF unit which would work with Boeing on the modernisation programme for almost four years. The RNZAF took delivery of the first

Orion NZ4201 and the No. 5 Squadron team which won the Fincastle Trophy in Cornwall in 1980. The trophy, competed for by maritime squadrons of Commonwealth air forces, was first won by the RNZAF in 1964. The 1980 team was captained by Squadron Leader R. M. Bulger (standing, second from right) and the officer commanding the squadron at the time was Wing Commander G. W. Ragg (kneeling at left).

Orion NZ4201, making a mock attack on a submarine during an exercise.

modified Orion in November 1983. The other four aircraft were modified in New Zealand, Air New Zealand acting as a sub-contractor to the Boeing Company, the last Orion being completed on schedule in May 1984.

The original radar in the Orion fleet had been designed in the early 1950s as a weather radar and was limited in the wider surveillance role, especially for the detection of small vessels in rough seas. The replacement radar, designed specifically for ocean surveillance, gave the Orions a much improved operational effectiveness.

Before modernisation the Orion crews had limited ability to identify vessels or observe ships' activities during the hours of darkness. They were also restricted in their search and rescue missions at night. The installation of an infra-red detection system, using equipment already in service in Australia, Canada and the United States, overcame these deficiencies. New inertial and Omega Navigation Systems were also installed, together with a data handling and display system.

When Phase Two was completed the RNZAF would have an up-to-date maritime surveillance and anti-submarine warfare capability which should remain effective into the next century.

A contract to purchase a sixth Orion at a cost of $NZ19 million was signed early in 1985, and Wing Commander M. M. Spring, Director of Communications-Electronics at Defence Headquarters, Wellington, who had administered the Orion update programme in New Zealand since mid 1980, took delivery of the aircraft at RAAF Base Edinburgh on 2 May. The aircraft, which had been in service with the Royal Australian Air Force, was bought through the Lockheed Corporation. After acceptance testing, Wing Commander R. M. Bulger, officer commanding No. 5 Squadron, and his crew flew it to New Zealand the following day. Modified by Air New Zealand on behalf of the Boeing Company, the Orion became operational in June 1986.

The Skyhawk Programme

When the RNZAF Skyhawks were purchased in 1968 they had no more than a general capability for close air support and undemanding interdiction tasks. This was acceptable when the requirement was for a combat force contribution operating in conjunction with allies, but the Government's Defence Review of 1983 confirmed the need for an expanded range of air combat tasks to defend New Zealand interests — maritime strike, counter-air, sea and land interdiction and close air support. The RNZAF, acknowledging that these tasks could be performed by a single aircraft type, was led to consider three options: the replacement of the A-4 with a more capable part-life (i.e. second-hand) aircraft; the acquisition of new aircraft; the retention and upgrading of the A-4. The part-life option was rejected on the grounds of cost-effectiveness; the cost of a new aircraft type would be prohibitive and so it was decided to upgrade the A-4.

While plans for upgrading were being examined, the RNZAF was given an opportunity to augment its force of strike aircraft. Available in Australia were a number of Skyhawks which had been in service with the Royal Australian Navy and were no longer needed. In 1984 the New Zealand Government approved the purchase of eight single seat A-4Gs and two TA-4G two-seat trainers. The aircraft were ferried to New Zealand by RNZAF pilots, the first four arriving at Ohakea on 12 July. The sum of

Former Australian naval Skyhawk NZ6213 undergoing modification at Woodbourne in 1985.

The contract with Lear Siegler International for updating the RNZAF Skyhawks was signed in March 1986. From left: the Secretary of Defence, Mr D. B. G. McLean, the Chief of the Air Staff, Air Vice-Marshal D. M. Crooks, the Minister of Defence, the Hon. F. O'Flynn, and the President of Lear Siegler's International Division, Mr C. S. Sikes.

$68.596 million had been earmarked for the purchase of the aircraft and associated spares, and to finance engineering refurbishment of the total Skyhawk fleet, including conversion of the A-4Gs to A-4K configuration, this being a major task for No. 1 Repair Depot at Woodbourne.

To accommodate the additional Skyhawks, the RNZAF relocated No. 42 Squadron at Auckland and, using the hangar, offices and other facilities made available at Ohakea, created an additional Strike Wing unit to be known as No. 2 Squadron. The last regular unit of that name — No. 2 (BR) Squadron — had been renamed No. 75 Squadron in October 1946 and the only subsequent unit to carry the name had been the Wellington squadron of the Territorial Air Force, from 1948 to 1957. Command of No. 2 Squadron was given to Squadron Leader J. S. Bates, AFC.

A proposal for the modernisation of the Skyhawk fleet's avionics system, using the latest in United States military technology, at a cost of $NZ140 million, was approved by the New Zealand Government in 1985. The modernisation programme included a new inertial navigation system using a ring laser gyro, modern multi-mode radar, a head-up display, fully computerised weapons management and aiming systems, the ability to launch and guide modern stand-off missiles, and a range of threat warning and self-defence devices. The contract, awarded to Lear

Siegler of New York in 1985, was to be completed within three years.

In an address to the Command and Staff College in December 1985, the Chief of Air Staff, Air Vice-Marshal D. M. Crooks, said that the upgraded A-4 would be a very formidable weapon, with attack and self-defence capabilities no less than modern aircraft such as the F-16 and F-18. It was specifically designed to meet New Zealand's defence requirements. No one pretended that the A-4 was an air superiority fighter, he said, but that was not the immediate requirement. 'It is our judgment that the threat of direct air attack against New Zealand or our regional interests is unlikely in the near future', he said. 'Beyond that we cannot say; certainly, the A-4's successor should have an air fighting capability whether or not a threat is then evident.'

In late 1985 the RNZAF was also examining the question of air refuelling for its combat force, primarily to extend its combat radius and provide for increased tactical flexibility in its deployment. This study was continued in 1986.

New Unit at Woodbourne

The relocation of No. 42 Squadron at RNZAF Base Auckland on 3 December 1984 to facilitate the extension of Strike Wing operations and the formation of No. 2 Squadron terminated almost 35 years of unbroken service at Ohakea for this unit. With this move, three Cessna 421C Golden Eagle aircraft which had been bought new in 1981 — enabling the last few de Havilland Devons to be phased out of service — were separated from 42 Squadron and transferred to RNZAF Base Woodbourne. This change gave Woodbourne its first complete flying unit in 40 years.

On 21 March 1985 Air Vice-Marshal Crooks was the Reviewing Officer for a formal wing parade at Woodbourne to mark the renaming of the Cessna Flight as No. 104 Flight. Crooks had long been interested in perpetuating the traditions of the Service and, encouraged by the reactivation of No. 3 Squadron in 1965, of No. 1 Squadron in 1972 and No. 2 Squadron in 1984, he would have favoured adding a No. 4 Squadron to the RNZAF's Order of Battle, thus perpetuating the four original squadrons. The three Cessnas, however, did not constitute a squadron, and as he did not favour descriptive titles (which often had unfortunate acronyms) he adopted a '100-number' series for smaller units and the Cessna Flight incorporated a number 4. At the parade on 21 March, he handed No. 4 Squadron's official badge, unit history and other memorabilia into the safe keeping of 104 Flight.

Himself a former adjutant of No. 4 Squadron in its TAF days, Air Vice-Marshal Crooks had as special guests that day Group Captain F. R. Dix and Squadron Leaders J. R. Day, N. B. Collins and E. Brooke-Taylor, former commanding officers of 4 Squadron, and Flight Lieutenant J. C. Hazledine, a former squadron member. In a similar move, RNZAF Support Unit Singapore was retitled No. 141 Flight on 30 April 1985, thus maintaining a reminder of the long and dedicated service of No. 41 Squadron.

A Cessna Golden Eagle of No. 42 Squadron, later of No. 104 Flight.

CHAPTER TWENTY-EIGHT

With the Multi-national Force and Observers in the Sinai

Late in 1981 the Royal New Zealand Air Force was allotted a new task which would see its members making a contribution to Middle East stability for the first time since No. 14 (Fighter) Squadron withdrew from Cyprus in 1955. In 1981 the Multi-national Force and Observers (MFO) of some 2600 men and women was formed from among 11 contributing nations to monitor the observance of the Egyptian/Israeli Peace Treaty, signed in March 1979.

Unlike earlier peacekeeping forces, the MFO was not a United Nations force. The Camp David accord* did not have the unanimous support of all the Arab countries, and if the concept of the MFO had been submitted to the United Nations Security Council there was every chance of the veto being applied. The effective date for the monitoring of the Egyptian/Israeli Peace Treaty was 25 April 1982 when all Israeli military personnel and civilians left the last zone in the Sinai Peninsula. For New Zealand, Anzac Day was a most appropriate date, for its main contribution was to be through an Anzac unit, made up of members of the Royal Australian Air Force, Royal Australian Navy, Royal New Zealand Air Force and New Zealand Army which, by agreement, would be commanded by an Australian. The Rotary Wing Aviation Unit (RWAU) was equipped with 10 Iroquois helicopters, two of which flew in RNZAF livery but were leased from the United States.

* The initial 'peace framework' agreement negotiated by US President Jimmy Carter, Egypt's President Anwar Sadat and Israel's Prime Minister Menachem Begin at Camp David, USA, on 17 September 1978.

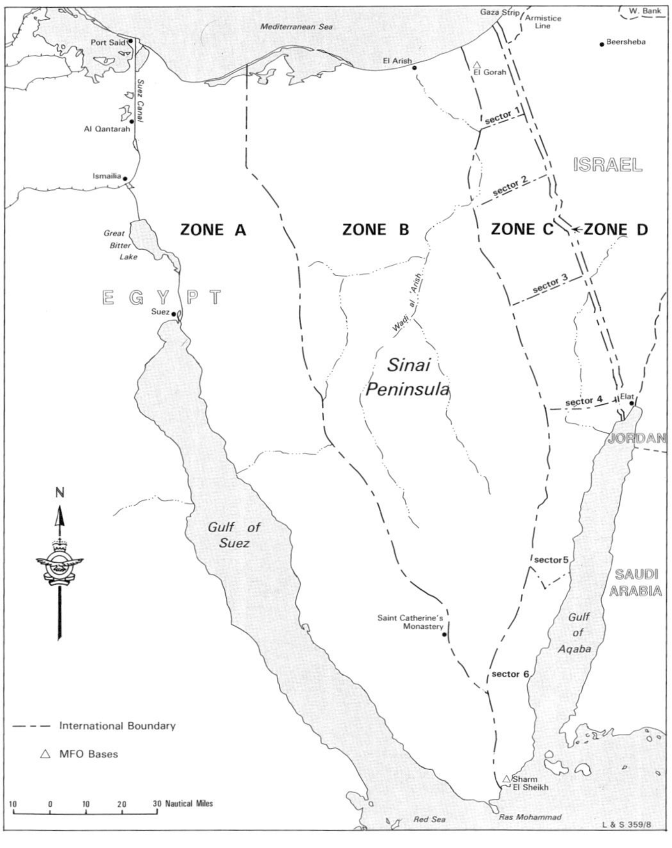

New Zealand's commitment to the MFO was 36 personnel — 29 from the Air Force and seven from the Army — and the two Iroquois aircraft. The only helicopters in the force other than those of the RWAU were those of a United States Army aviation company operating in the southern Sinai.

Under the protocol of the treaty the Sinai Peninsula was divided into zones from which the Israelis withdrew in phases. The so-called 'buffer zone' (Zone C), in which the MFO was the only military presence, extended in a narrow 500 kilometre strip on the Egyptian side of the international border from the Mediterranean coast in the north to Sharm-el-Sheik on the Gulf of Aquaba in the south. The Egyptians occupied Zones A and B to the west while the Israelis occupied Zone D, a 3 kilometre ribbon of land extending along the Israeli side of the international border. The number of armed forces permitted in each zone was limited by the terms of the Peace Treaty.

The RNZAF contribution included four Iroquois pilots and two crewmen in the RWAU, and seven aircraft maintenance technicians, besides communications staff and administrative personnel. Wing Commander N. J. S. Rodger was the first RNZAF staff officer in the Force Headquarters and the first New Zealand contingent commander. He travelled ahead of his party in April 1982 and Squadron Leader G. L. Wood escorted the balance of the New Zealand contingent to their new home at El Gorah in an RAAF Boeing 707 transport aircraft.

The New Zealand contingent commander had a dual role in that he was not only responsible for the New Zealanders with the MFO but also assisted the Chief of Staff, a senior US Army officer, in the operation of the force's mission in the Sinai — the prevention of any violation of the Peace Treaty.

El Gorah, previously known as Etam, was a former Israeli air base in north-east Sinai which had been dismantled by the Israelis as they withdrew; the MFO had to rebuild the facilities. Here the New Zealanders of the

RWAU settled down to a completely new life. The unit's tasks were varied. The Iroquois crews provided air transport and logistic support for the Colombian and Fijian infantry battalions manning remote observation posts and checkpoints in the central and northern regions of the zone. They also supported civilian observers from the US State Department, flying them out into all four zones to monitor Egyptian and Israeli forces and their weaponry. They maintained a 24-hour search and rescue standby service and undertook medical evacuation flights, most of which were to El Gorah, El Arish or Tel Aviv.

The unit was also responsible for continuation training and theatre indoctrination as replacements arrived. The resupply of personnel and equipment was generally handled by C-130 aircraft of the RNZAF and RAAF every six weeks. The first flight into the Sinai by an RNZAF Boeing 727 of No. 40 Squadron was made in November 1984.

While the RWAU absorbed the greater part of the New Zealand contribution to the MFO, seven New Zealanders — two from the RNZAF and five from the New Zealand Army — served 12-month tours on the staff of the Force Headquarters at El Gorah. Personnel of the helicopter unit served for periods of six months, living and working in an environment that was generally harsh by New Zealand standards. Summertime temperatures would reach a blistering 45°C (although the average temperature would be about 38°C) and winter would bring a sharp drop to 12°C with heavy rain, hail and even snow at times. For the pilots the featureless terrain, the inadequacy of navigational aids and dust storms of extraordinary severity and prolonged duration — rising sometimes to an altitude of 10,000 feet — were among the major hazards.

The accommodation was rather cramped in the first few months of the mission but facilities tended to develop around the MFO and in time the force built new barracks. A gymnasium, swimming pool and tennis courts added later made for an improved morale. The RWAU built its own all-ranks club — the Anzac Surf Club — which was very popular with all contingents. Although the Anzac nature of the helicopter unit was genuinely maintained and even fostered (the Iroquois would normally operate with mixed crews), the contributing countries were nevertheless strongly nationalistic. They flew their own national flags, for example, and the New Zealanders' contribution to MFO entertainment usually included a Maori concert party or at least a haka group.

At the controls of an Iroquois helicopter.

In 1983 the New Zealanders joined with the Australians to raise more than $5,000 for the Australian Bushfire Fund and later staged a 500 kilometre run through the length of the Sinai Peninsula in blistering temperatures to raise $15,000 for New Zealand's Telethon appeal of that year.

The RNZAF officers who succeeded Wing Commander

Iroquois helicopters of the Rotary Wing Aviation Unit over one of the pyramids in 1986.

Rodger as New Zealand Contingent Commander — Wing Commanders T. H. J. Knight, I. G. Brunton and T. W. Medcalf — found that same spirit of Anzac co-operation very much in evidence.

The first RNZAF women to go there were Sergeant Gayle Chambers and Corporal Joyce McGee who were posted to the MFO in January 1985, the former to work in personnel administration and the latter in the operations centre as a telecommunications operator.

By the end of March 1986, when New Zealand and Australian participation in the scheme came to an end, about 200 RNZAF men and women had completed tours of duty with the MFO in the Sinai.

An RNZAF Skyhawk over Wellington Airport.

CHAPTER TWENTY-NINE

Perspectives and Prospects

After 50 years of service as an independent branch of the New Zealand Armed Forces, the Royal New Zealand Air Force is of an age to look back with pride. Belonging to a small nation with limited financial resources, it has never been the recipient of indulgence; it has often had to work hard to convince the bureaucrats of its needs and in peacetime it has frequently had too few aircraft for the tasks asked of it. That it has acquitted itself with distinction over five decades is in itself a source of satisfaction for those who have served in its ranks.

Through those 50 years of endeavour the RNZAF has acquired a maturity, a sophistication, a confidence in its ability and a modest awareness of its own worth. Like the Regular serviceman who enlists as a raw recruit and accumulates layers of knowledge and experience in a lifetime career, so the RNZAF as an instrument of national defence has progressed from early flying machines of supreme simplicity to aircraft of extreme complexity, with weapons systems to match, and in the process has become a highly professional organisation. In the words of Air Vice-Marshal D. M. Crooks, today's serviceman is better educated, better trained and better motivated than his counterpart of 50 years ago 'because he has to be'. This opinion is readily confirmed by a study of any one squadron, maintenance unit or training establishment where the required standards are uniformly exacting.

Over the years New Zealand has loosened her once firm ties with Britain, politically, socially and economically. As our country has, perhaps imperceptibly, developed its own national character so has the RNZAF shed much of the RAF image that marked its first two decades, although its heritage as a Service that was moulded in its formative years by RAF officers and senior NCOs is still clearly apparent. Despite the fact that its squadrons have served and exercised alongside the USAF, the RAAF and the RAF, it is the latter's influence that is still most evident in the RNZAF today. Its uniform, although it has been modified, clearly derives from the RAF, as does its rank structure, its organisation and its basic philosophy relating to concepts of air power and the roles, tasks and functions of an air force. This same heritage is apparent in the RNZAF's ceremonial; the presentation of the Queen's Colour in 1953 and of Standards to the operational squadrons since then have all reflected the British origin of these traditions. Nevertheless, the RNZAF has developed a recognisable national character.

The roles of the Central Band of the RNZAF and the various base bands also reflect a British origin. The Central Band — now the Service's sole remaining territorial unit — is more than 50 years old, having been established within the Territorial Air Force as long ago as 1935. Since its debut on 12 May 1937, when it led the procession through New Zealand's capital city of Wellington for the Coronation of King George VI, the band has seen many changes to its form and membership, but its high standards of musicianship have been well maintained. Like the base bands, it has lent colour, dignity and precision to major ceremonial events in the life of the RNZAF and, like the Queen's Colour and the Squadron Standards, its existence as an integral part of an Armed Service is part of an ancient tradition.

With only a half century of service behind it the RNZAF is already rich in tradition, and those men and women who today wear its uniform are conscious of its contribution to military history. Its participation in every major area of conflict during the Second World War alone ensures it a lasting place in history, but its contributions since then have been notable, as this book has sought to show.

In the RNZAF Museum at Wigram there are plenty of reminders of its past. Displays of early photographs depict the pioneering spirit and enterprise of New Zealand's first aviators; portraits of men like Sir Henry Wigram, Sergeant J. A. Ward, Wing Commander L. H. Trent,

Flight Lieutenant R. H. Simpson, MBE, who retired as Director of Music in 1971.

Squadron Leader G. H. P. Hanify, MBE, appointed Director of Music in 1972.

Flying Officer L. A. Trigg and Group Captain D. J. Scott bear witness that vision, drive and generosity of heart went hand in hand with leadership by example and courage in the air in creating an air force second to none in fighting spirit, initiative and airmanship.

Some early aircraft types preserved in the museum recall wartime and peacetime roles and special events — Dakota NZ3551, the pride of No. 42 Squadron when it carried Royalty; Grumman Avenger NZ2504, restored after 2000 hours of labour, a reminder of the Northern Solomons campaign and of postwar aerial topdressing experiments; and the Auster monoplane which Wing Commander J. R. Claydon and his flight operated in Antarctica in 1956. The Australian gift Canberra, the Mustang fighter acquired from the Indonesian Air Force, a beautifully restored Spitfire, a Vampire FB.5, a Devon navigation trainer, and a Bristol Freighter evoke memories of different phases in the 50-year history of the RNZAF. A Vildebeest, a Hudson, a Catalina, an Oxford and an Anson were among others in various stages of restoration in 1986. A Harvard, an Avro 626 and a Tiger Moth were being maintained in flying condition with the Historic Flight of the Central Flying School at Wigram.

The museum, whose patron is the Prince of Wales, will do more than provide a permanent home for aircraft, engines, armament, trophies and other exhibits of historical significance. It will also assist in training young men and women in the history of their Service and inspire and motivate them by a study of its achievements and traditions.

Over the first 50 years of its operations as an independent Service, the RNZAF has developed a significant degree of self-reliance and in some areas, notably training, is almost self-sufficient. The standards of RNZAF training, air and ground, have long been respected, not only by other Armed Forces but also by industry. From long-established traditions RNZAF training schools have, over the years, developed standards of excellence that bear comparison anywhere in the world. This is one of the great strengths of the RNZAF. In the post-1987 years these schools may need to expand their scope.

These and other considerations have led to the concept of an Air Force College where the various elements of training and education can be drawn together at one location with modern facilities. Wigram has been selected as the logical location. It is already the hub of officer training; it has ties with Canterbury University where a successful University Cadet Scheme for the RNZAF has been in operation for more than 20 years; and it will be the site of the Air Force's principal library, archives and research facilities in conjunction with the developing museum.

In 1986 the RNZAF with a strength of 4176 — 638 officers, 79 officer cadets, and 3459 other ranks — was little bigger than the Regular air force of the 1950s, but it was well trained and efficient, with units working hard to meet their commitments. Some were affected by staff shortages, many aircrew members and senior aircraft technicians having left the Service before their engagements expired, attracted to civilian industry by the promise of better financial rewards.

In the Air Force that year there were many concerns — about the future of defence, the direction the country was taking, and the impact of the Government's anti-nuclear policy on relations with the US Government and the Anzus Pact of 1951 in particular. The whole question of defence was being subjected to a closer public and political scrutiny, with a concomitant questioning of traditional values. There was little doubt that New Zealand had reached a crossroads, both in her perceptions of national security and the future course of her international relationships.

But one thing is certain — whatever the changes in defence policy, in military alignments, and in regional emphasis, there will always be a primary role for the RNZAF. The air force which Cochrane shaped, which Saunders expanded and developed, and which successive chiefs of staff have nurtured, refined and steered towards maturity, has emerged in the late 1980s as a highly professional body of men and women. There might be new directions ahead for the country and its Armed Forces, but the RNZAF will continue to meet its obligations and its pursuit of them should be challenging and exciting.

Skyhawks of No. 75 Squadron over Penang, West Malaysia, exercising with aircraft of the Royal Malaysian Air Force.

A Hercules transport of No. 40 Squadron on the ice runway at McMurdo Sound, Antarctica.

Appendix A

Squadrons of the RNZAF

NO. 1 SQUADRON

Formed at Auckland on 1 August 1930 as No. 1 (Army Co-operation) Squadron.

Sqn Ldr J. Seabrook, AFC	Aug 1930
Sqn Ldr J. D. Hewett	Mar 1934

Renamed Auckland Territorial Squadron September 1938.

Sqn Ldr D. M. Allan	Sep 1938
Sqn Ldr C. L. Monckton	Sep 1939
Sqn Ldr W. G. Coull	Sep 1939

Renamed New Zealand General Reconnaissance Squadron March 1940.

Sqn Ldr G. N. Roberts, AFC	Mar 1940

Renamed No. 1 (General Reconnaissance) Squadron, January 1941. (Title later changed to No. 1 (Bomber Reconnaissance) Squadron.)

Sqn Ldr G. N. Roberts, AFC	Jan 1941
Sqn Ldr G. H. Fisher	Aug 1941
Sqn Ldr F. J. Lucas, DFC (and bar)	Jul 1942
Sqn Ldr C. L. Monckton	Dec 1942
Sqn Ldr E. W. Tacon, DFC, AFC	Mar 1943
Wg Cdr H. C. Walker, AFC	May 1943
Sqn Ldr K. G. King	Apr 1944
Wg Cdr A. N. Johnstone	May 1944
Wg Cdr A. A. N. Breckon, DFC	Feb 1945

Disbanded 10 September 1945.

Reformed at Auckland on 1 December 1948 as No. 1 (Auckland) Territorial Squadron.

Sqn Ldr W. K. Paterson, AFC	Dec 1948
Sqn Ldr J. R. Court, DFC	Dec 1951
Sqn Ldr A. A. Watson, DFC	Jan 1955

Disbanded 1 August 1957.

Reformed at Whenuapai on 1 August, 1972 as No. 1 Squadron.

Sqn Ldr P. R. Adamson	Aug 1972
Sqn Ldr C. F. L. Jenks	Dec 1973
Sqn Ldr G. A. Oldfield	Apr 1976
Sqn Ldr R. S. Holdaway	Aug 1976
Sqn Ldr G. A. Oldfield	Feb 1977
Sqn Ldr K. A. Skilling	Sep 1977
Sqn Ldr R. S. Holdaway	Feb 1978
Sqn Ldr K. J. Wells, DFC	Jan 1980
Sqn Ldr I. W. Collins	Jan 1982
Sqn Ldr K. L. Crofskey	Jul 1984

Disestablished 7 December 1984.

NO. 2 SQUADRON

Formed in Wellington on 1 August 1930 as No. 2 (Bomber) Squadron.

Sqn Ldr M. C. McGregor, DFC (and bar)	Aug 1930
Sqn Ldr G. L. Stedman, ED	Jul 1931
Sqn Ldr E. A. Gibson	Aug 1937

Renamed Wellington Territorial Squadron November 1937.

Sqn Ldr E. A. Gibson	Nov 1937
Sqn Ldr R. J. Sinclair	Sep 1939

Disbanded on formation of New Zealand General Reconnaissance Squadron.

Reformed at Nelson on 1 January 1941 as No. 2 (General Reconnaissance) Squadron. (Title later changed to No. 2 (Bomber Reconnaissance) Squadron.)

Sqn Ldr R. J. Cohen, AFC	Jan 1941
Sqn Ldr J. J. Busch	May 1942
Sqn Ldr E. A. A. Moen, AFC	Jan 1943
Sqn Ldr R. J. K. Hogg, DFC, RAF	Apr 1943
Wg Cdr A. B. Greenaway	Feb 1944
Wg Cdr W. A. Cameron	June 1944
Wg Cdr K. G. King, AFC	May 1945
Wg Cdr R. F. Watson, AFC	Oct 1945
Wg Cdr A. B. Greenaway	Oct 1945

Renamed No. 75 Squadron 11 October 1946.

Reformed at Wellington on 1 December 1948 as No. 2 (Wellington) Territorial Squadron.

Sqn Ldr J. A. Oldfield, DFC	Dec 1948
Flt Lt R. G. Jeffs (temporary)	Dec 1951
Sqn Ldr M. B. Stevens	Jan 1952
Sqn Ldr E. D. McCabe	Jan 1954

Disbanded 1 August 1957.

Reformed at Ohakea on 11 December 1984 as No. 2 Squadron.

Sqn Ldr J. S. Bates, AFC	Dec 1984
Sqn Ldr J. R. Jennings	Sept 1985

NO. 3 SQUADRON

Formed at Christchurch on 8 August 1930 as No. 3 (Bomber) Squadron.

Sqn Ldr N. E. Chandler	Aug 1930
Sqn Ldr W. L. Harrison	Dec 1931

Renamed Christchurch Territorial Squadron September 1938.

Sqn Ldr G. L. Stedman, ED	Sept 1938

Disbanded on formation of New Zealand General Reconnaissance Squadron March 1940.

Reformed at Harewood on 1 April 1941 as No. 3 (General Reconnaissance) Squadron. (Title later changed to No. 3 (Bomber Reconnaissance) Squadron.)

Sqn Ldr C. L. Monckton	Apr 1941
Sqn Ldr J. W. H. Bray	May 1942
Wg Cdr G. H. Fisher	Sep 1942*
Sqn Ldr D. E. Grigg, MBE	Apr 1943
Sqn Ldr J. J. Busch	July 1943
Sqn Ldr A. N. Johnstone	Nov 1943
Sqn Ldr A. B. Greenaway	Dec 1943†
Wg Cdr I. G. Morrison	Feb 1944

Disbanded 25 June 1945.

Reformed at Christchurch on 1 December 1948 as No. 3 (Canterbury) Territorial Squadron.

Sqn Ldr J. R. Maling, AFC	Dec 1948
Sqn Ldr R. J. N. Archibald, DFC	Jan 1952
Flt Lt M. F. McD. Palmer	Jan 1956

* Squadron Leader H. C. Walker, AFC, commanded the RNZAF detachment at Guadalcanal from January to April 1943 and was awarded the US Legion of Merit for duties as CO of No. 3 Squadron during this period. Documents held in National Archives (Air Accident Files) show that Wing Commander G. H. Fisher, who was based at rear echelon Espiritu Santo, was CO No. 3 Squadron until April 1943. The squadron's forward echelon was at Guadalcanal.

† Squadron Leader Johnstone left Espiritu Santo on 10 December 1943 and Squadron Leader A. B. Greenaway (Officer Commanding No. 2 Squadron) commanded both squadrons until No. 3 Squadron personnel returned to New Zealand on 11 January 1944.

Flt Lt H. E. Tully	Feb 1957

Disbanded 1 August 1957.
Reformed at Hobsonville on 1 August 1965 as No. 3 (Battlefield Support) Squadron.

Sqn Ldr I. A. Hutchins, AFC (acting)	May 1965
Sqn Ldr W. J. Cranfield, AFC (acting)	Aug 1965
Wg Cdr C. L. Siegert, MVO, DFC, AFC	Jan 1966
Wg Cdr D. E. Jamieson, OBE	Nov 1966
Wg Cdr D. M. Dallison	Nov 1967
Wg Cdr C. W. Rudd, AFC	Dec 1969

Renamed No. 3 Squadron 1 May 1970.

Wg Cdr C. W. Rudd, AFC	May 1970
Wg Cdr G. C. Derby	Jan 1972
Wg Cdr W. J. Cranfield, AFC	Jul 1974
Wg Cdr R. J. Klitscher, DFC	Jul 1976
Wg Cdr R. S. Holdaway	Jan 1980
Wg Cdr K. J. Wells, DFC	Jan 1982
Wg Cdr P. J. Garnett, MBE	Jun 1983
Wg Cdr G. L. Wood	Dec 1985

NO. 4 SQUADRON
Formed at Dunedin on 8 August 1930 as No. 4 (Army Co-operation) Squadron.

Sqn Ldr P. K. Fowler	Aug 1930
Sqn Ldr T. W. White	Mar 1932

Squadron personnel posted to Unattached List 15 September 1938 pending formation of a Dunedin territorial squadron. (The squadron was not formed as facilities and equipment were not available at that time.)
Reformed in Fiji on 8 October 1941 as No. 4 (General Reconnaissance) Squadron. (Title later changed to No. 4 (Bomber Reconnaissance) Squadron.)

Sqn Ldr G. R. White	Oct 1941
Sqn Ldr H. C. Walker	Nov 1942
Wg Cdr E. M. Lewis	Jan 1943
Sqn Ldr E. W. Tacon, DFC, AFC, RAF	May 1943
Sqn Ldr E. Brooke-Taylor	Nov 1943
Sqn Ldr N. R. Lecher	Apr 1944
Wg Cdr F. R. Dix	Jan 1945
Wg Cdr A. F. H. Tye, DFC	Mar 1945
Wg Cdr L. H. Parry	Jul 1945
Sqn Ldr G. S. A. Stevenson	Sep 1945

Disbanded September 1945.
Reformed at Dunedin on 1 December 1948 as No. 4 (Otago) Territorial Squadron.

Sqn Ldr W. R. Kofoed, DSO, DFC	Dec 1948
Sqn Ldr J. R. Day, MBE, AFC	Dec 1950
Sqn Ldr G. C. N. Johnson	Dec 1952
Sqn Ldr J. R. Day, MBE, AFC	Jan 1953
Sqn Ldr N. B. Collins	Jun 1954

Disbanded 1 August 1957.

NO. 5 SQUADRON
Formed in Fiji on 18 November 1941 as No. 5 (General Reconnaissance) Squadron. (Title later changed to No. 5 (Bomber Reconnaissance) Squadron.)

Sqn Ldr E. M. Lewis	Nov 1941
Sqn Ldr W. J. Craig, RAF	Dec 1941
Flt Lt R. H. F. Hickson	Feb 1942
Sqn Ldr R. J. R. H. Makgill	Apr 1942
Flt Lt I. R. Salmond	Aug 1942
Flt Lt G. K. D. Mackenzie	Nov 1942

Disbanded 20 November 1942.
Reformed at Fiji on 10 July 1944 as No. 5 (Flying Boat) Squadron. (Later No. 5 (Maritime) Squadron, then No. 5 Squadron.)

Wg Cdr J. W. H. Bray	Jul 1944
Wg Cdr J. M. Checketts, DSO, DFC	Nov 1947
Sqn Ldr C. R. Hutton	Dec 1949
Sqn Ldr S. O. Field	Jul 1951
Sqn Ldr J. B. Wilson	Jul 1953
Wg Cdr R. K. Walker, AFC	Apr 1954
Wg Cdr R. D. McVicker, AFC	Oct 1956
Wg Cdr R. L. Scott, DFC	May 1958
Wg Cdr J. L. Scott, DFC	Oct 1961
Wg Cdr D. B. Flintoff	Jan 1964
Sqn Ldr W. H. Willis	Feb 1965
Sqn Ldr B. J. Dwyer (acting)	June 1965
Wg Cdr D. B. Flintoff	Oct 1965
Sqn Ldr K. McL. Milne	Apr 1966
Flt Lt J. A. Laing (acting)	Mar 1967
Wg Cdr D. B. Flintoff	Apr 1967
Sqn Ldr N. E. Richardson	Apr 1968
Cdr P. C. Davenport, US Navy	Jul 1968
Wg Wdr N. E. Richardson	Jan 1969
Sqn Ldr P. J. McKay (acting)	Dec 1969
Wg Cdr I. V. Mackay	Jan 1970
Wg Cdr P. J. McKay	Jan 1973
Wg Cdr D. W. Pawson, AFC	Jun 1975
Wg Cdr G. W. Ragg	Jul 1978
Wg Cdr D. G. Simpson	Jul 1981
Wg Cdr R. M. Bulger	Jul 1984
Wg Cdr T. W. Medcalf	Feb 1986

NO. 6 SQUADRON
Formed at Milson on 20 February 1942 as No. 6 (Army Co-operation) Squadron.

Wg Cdr J. G. Fraser, RAF*	Feb 1942

Disbanded 20 October 1942 on formation of No. 21 (Army Co-operation) Squadron.
Reformed at Fiji on 25 May 1943 as No. 6 (Flying-Boat) Squadron.

Wg Cdr G. G. Stead, RAF	May 1943
Wg Cdr I. A. Scott	Aug 1943
Wg Cdr J. R. S. Agar	Aug 1944
Wg Cdr K. G. Smith	Jun 1945

Disbanded 31 August 1945.
Reformed at Hobsonville on 2 April 1952 as No. 6 (Maritime) Squadron.

Flt Lt D. F. Clarke, DFC, AFC	Apr 1952
Sqn Ldr R. L. Scott, DFC	May 1952
Flt Lt D. F. Clarke, DFC, AFC	Aug 1952
Sqn Ldr R. L. Scott, DFC	Aug 1953
Sqn Ldr R. K. Walker, AFC	Jan 1954
Sqn Ldr J. B. Wilson	Apr 1954
Sqn Ldr D. T. Sheehan	May 1955
Sqn Ldr J. B. Spencer	Jan 1956

Disbanded 1 August 1957.

* Also Commanding Officer No. 20 (AC) Squadron.

NO. 7 SQUADRON
Formed at Waipapakauri on 16 February 1942 as No. 7 (General Reconnaissance) Squadron.

Wg Cdr A. J. Turner	Feb 1942
Wg Cdr R. A. Kirkup	Dec 1942
Wg Cdr R. J. Sinclair	May 1943

Disbanded 27 May 1943.

NO. 8 SQUADRON
Formed at Whenuapai on 22 April 1942 as No. 8 (General Reconnaissance) Squadron.

Sqn Ldr C. L. Monckton	Apr 1942
Sqn Ldr I. G. Morrison	Sep 1942
Sqn Ldr M. Wilkes	Apr 1943
Flt Lt F. J. Adams	May 1943

Disbanded 28 May 1943 and reformed as No. 30 Squadron.
Reformed at Whenuapai on 1 October 1944 as No. 8 (Bomber Reconnaissance) Squadron.

Wg Cdr L. H. Parry	Oct 1944
Wg Cdr I. R. Salmond	Nov 1944

Disbanded 12 March 1945.

NO. 9 SQUADRON
Formed in New Caledonia on 23 July 1942 as No. 9 (General Reconnaissance) Squadron. (Title later changed to No. 9 (Bomber Reconnaissance) Squadron.)

Sqn Ldr D. E. Grigg, MBE	Jul 1942
Sqn Ldr J. J. Busch	Mar 1943
Sqn Ldr A. E. Willis	Jul 1943
Wg Cdr A. C. Allen, AFC	Oct 1943

Disbanded 25 June 1945.

NO. 14 SQUADRON
Formed at Ohakea and Masterton on 25 April 1942 as No 14 (Fighter) Squadron.* (Word 'Fighter' later deleted from title.)

Sqn Ldr J. N. MacKenzie, DFC, RAF	Apr 1942
Sqn Ldr R. E. P. Brooker, DFC (and bar), RAF	Jul 1942
Sqn Ldr T. B. Fitzgerald, DFC, RAF	Feb 1943
Sqn Ldr S. G. Quill	May 1943
Flt Lt H. R. Wigley (acting)	July 1943
Sqn Ldr S. G. Quill	Sep 1943
Sqn Ldr J. C. Blundell	Apr 1944
Sqn Ldr D. W. Cocks	Jul 1944
Sqn Ldr P. R. McNab	Feb 1945
Sqn Ldr J. R. C. Kilian	Jun 1945
Sqn Ldr J. J. de Willimoff, MBE, DFC	Dec 1945
Sqn Ldr D. F. St George, DFC	Apr 1947
Sqn Ldr D. S. N. Constance, DFC	Jan 1949
Sqn Ldr L. A. Rayner, DFC	Aug 1951
Sqn Ldr S. M. Hope, MBE, AFC	Oct 1951
Sqn Ldr N. H. Bright	May 1954
Sqn Ldr A. F. Tucker, DFC	Sep 1956
Sqn Ldr H. G. Moss	Jan 1959
Sqn Ldr M. F. McD. Palmer	Jan 1961
Sqn Ldr S. McIntyre, DFC	Apr 1962
Sqn Ldr G. Wallingford	Jun 1964
Sqn Ldr P. Neville, AFC	Dec 1966
Sqn Ldr D. R. Smith	Apr 1969
Sqn Ldr J. S. Boys, AFC	Sep 1970
Sqn Ldr J. S. Hosie	Jun 1972
Sqn Ldr R. F. Lawry, AFC	Jan 1974
Sqn Ldr L. A. Olsen, AFC	Jan 1975
Sqn Ldr P. B. Curtin, AFC	Jun 1978
Sqn Ldr J. W. Lanham	Feb 1979
Sqn Ldr P. S. Faulkner (acting)	Nov 1980
Sqn Ldr K. A. Jones	Jan 1981
Sqn Ldr J. S. Barclay	Feb 1982
Sqn Ldr D. S. Glendinning	Jun 1983
Sqn Ldr I. G. Wood	Jul 1985
Sqn Ldr J. C. Grocott, RAF	Jun 1986
Sqn Ldr I. G. Wood	Dec 1986

* Initially known as No. 488 Squadron (until early June 1942).

NO. 15 SQUADRON
Formed at Whenuapai on 1 June 1942 as No. 15 (Fighter) Squadron.

Sqn Ldr A. Crichton	Jun 1942
Sqn Ldr M. J. Herrick, DFC, RAF	Mar 1943
Sqn Ldr J. A. A. Gibson, DFC, RAF	Dec 1943
Sqn Ldr D. P. Winstone	Aug 1944
Sqn Ldr M. R. Clarke	Jan 1945

Disbanded October 1945.

NO. 16 SQUADRON
Formed at Ohakea on 1 July 1942 as No. 16 (Fighter) Squadron.

Sqn Ldr A. N. Jones	Jul 1942
Sqn Ldr J. S. Nelson	May 1943
Sqn Ldr J. H. Arkwright	Oct 1943
Sqn Ldr A. G. Sievers	Jan 1944
Sqn Ldr M. C. P. Jones	Jun 1944
Sqn Ldr P. S. Green	Oct 1944
Sqn Ldr J. H. Mills	Jul 1945

Disbanded October 1945.

NO. 17 SQUADRON
Formed at Whenuapai on 10 October 1942 as No. 17 (Fighter) Squadron.

Sqn Ldr J. V. A. Reid	Oct 1942
Sqn Ldr P. G. H. Newton	Jul 1943
Sqn Ldr D. F. St George	Apr 1944
Sqn Ldr B. V. Le Pine	Nov 1944

Disbanded 26 September 1945.

NO. 18 SQUADRON
Formed at Woodbourne on 5 June 1943 as No. 18 (Fighter) Squadron.

Sqn Ldr J. A. Oldfield, DFC	Jun 1943
Sqn Ldr R. H. Balfour, DFC	May 1944
Sqn Ldr P. R. McNab	Aug 1944
Sqn Ldr G. H. Corbet	Nov 1944
Sqn Ldr W. A. Hardham	May 1945

Disbanded 8 October 1945.

NO. 19 SQUADRON
Formed at Ardmore on 10 December 1943 as No. 19 (Fighter) Squadron.

Sqn Ldr H. R. Wigley	Dec 1943

Sqn Ldr M. T. Vanderpump, DFC — May 1944
Sqn Ldr J. R. C. Kilian — Jul 1944
Sqn Ldr H. A. Eaton — Jun 1945
Disbanded 12 October 1945.

NO. 20 SQUADRON
Formed at Onerahi on 10 August 1942 as No. 20 (Army Co-operation) Squadron.
Wg Cdr J. G. Fraser, RAF* — Aug 1942
Flt Lt F. L. Gates, RAF — Mar 1943
Flt Lt F. Edwards — May 1943
Disbanded 1 July 1943.
Reformed at Ardmore on 25 January 1944 as No. 20 (Fighter) Squadron.
Sqn Ldr S. R. Duncan — Jan 1944
Sqn Ldr G. M. Robertson — Jul 1944
Sqn Ldr L. R. Reynolds — Dec 1944
Disbanded 10 September 1945.

NO. 21 SQUADRON
Formed at Ohakea on 20 October 1942 as No. 21 (Army Co-operation) Squadron.
Sqn Ldr S. N. Wiltshire, GC — Oct 1942
Sqn Ldr W. A. Cameron — Mar 1943
Disbanded 26 August 1943.
Reformed at Ardmore on 1 May 1944 as No. 21 (Fighter) Squadron.
Sqn Ldr C. R. Bush, DFC, RAF — May 1944
Sqn Ldr W. J. MacLeod — Mar 1945
Disbanded 10 September 1945.

NO. 22 SQUADRON
Formed at Norwood on 9 November 1942 as No. 22 (Army Co-operation) Squadron.
Sqn Ldr W. A. Cameron — Nov 1942
Sqn Ldr W. L. Harrison — Mar 1943
Disbanded 10 May 1943.
Reformed at Ardmore on 19 June 1944 as No. 22 (Fighter) Squadron.
Sqn Ldr B. H. Thomson — Jun 1944
Sqn Ldr J. R. Court — Mar 1945
Sqn Ldr G. A. Delves — Aug 1945
Disbanded 10 September 1945.

NO. 23 SQUADRON
Formed at Ardmore on 7 August 1944 as No. 23 (Fighter) Squadron.
Sqn Ldr J. J. de Willimoff, DFC — Aug 1944
Sqn Ldr D. E. Hogan, DFC — May 1945
Disbanded 15 October 1945.

NO. 24 SQUADRON
Formed at Ardmore on 22 September 1944 as No. 24 (Fighter) Squadron.
Sqn Ldr M. T. Vanderpump, DFC — Sep 1944
Sqn Ldr A. G. S. George, DFC — Feb 1945
Disbanded 31 October 1945.

NO. 25 SQUADRON
Formed at Seagrove on 31 July 1943 as No. 25 (Dive-Bomber) Squadron.

* Also Commanding Officer No. 6 (AC) Squadron February–October 1942.

Sqn Ldr T. J. MacLean de Lange — Jul 1943
Disbanded 19 June 1944.
Reformed at Ardmore on 30 October 1944 as No. 25 (Fighter) Squadron.
Sqn Ldr G. M. Fitzwater — Oct 1944
Disbanded 10 September 1945.

NO. 26 SQUADRON
Formed at Ardmore on 1 March 1945 as No. 26 (Fighter) Squadron.
Sqn Ldr G. A. Delves — Mar 1945
Disbanded 30 July 1945.

NO. 30 SQUADRON
Formed at Gisborne on 4 June 1943 as No. 30 (Dive-Bomber) Squadron.
Sqn Ldr R. G. Hartshorn — Jun 1943
Sqn Ldr H. N. James — April 1944
Disbanded 26 June 1944.

NO. 31 SQUADRON
Formed at Gisborne on 7 December 1943 as No. 31 (Dive-Bomber) Squadron.
Sqn Ldr M. Wilkes
Disbanded 1 September 1944.

NO. 40 SQUADRON
Formed at Whenuapai on 1 June 1943 as No. 40 (Transport) Squadron.
Sqn Ldr F. J. Lucas, DFC (and bar) — Jun 1943
Wg Cdr J. Adams, DFC, AFC — Jun 1943
Wg Cdr R. A. Kirkup — Nov 1944
Sqn Ldr M. L. Pirie, MBE — Apr 1946
Disbanded 31 October 1947.
Reformed at Whenuapai on 8 December 1954 as No. 40 Squadron.
Sqn Ldr R. B. Bolt, DFC — Dec 1954
Sqn Ldr K. B. Smith — Jun 1955
Sqn Ldr O. D. Staple, DFC — Nov 1957
Sqn Ldr G. H. S. Tosland, AFC — Sep 1959
Wg Cdr J. L. Trolove — Jan 1962
Sqn Ldr K. A. Sawyer, DFC — Dec 1962
Wg Cdr B. A. Wood — Nov 1963
Sqn Ldr K. A. Sawyer, DFC — Nov 1964
Wg Cdr B. A. Wood, AFC — Apr 1965
Wg Cdr B. J. O'Connor, AFC — Dec 1966
Wg Cdr M. W. Hodge — Nov 1969
Wg Cdr A. E. Thomson — Jan 1974
Wg Cdr M. E. A. Dillon — Jul 1976
Wg Cdr K. A. Gayfer — Apr 1979
Wg Cdr C. W. Adamson — Jul 1982
Wg Cdr T. K. Butler, DFC — Mar 1984
Wg Cdr A. T. King — Jan 1987

NO. 41 SQUADRON
Formed at Whenuapai on 15 August 1944 as No. 41 (Transport) Squadron. (Word 'Transport' later deleted from title.)
Sqn Ldr A. F. H. Tye, DFC (acting) — Aug 1944
Wg Cdr H. C. Walker, AFC — Sep 1944
Wg Cdr L. H. Parry, AFC — Sep 1945

Wg Cdr T. J. MacLean de Lange, DFC	Jun 1947
Sqn Ldr A. H. Harding, DFC	Mar 1949
Sqn Ldr J. C. Murray	Dec 1949
Wg Cdr R. F. Watson, AFC	Nov 1951
Wg Cdr A. P. Gainsford, DSO, DFC, AFC	May 1954
Sqn Ldr R. A. Manners, MBE	Dec 1954
Sqn Ldr A. S. Tie	Nov 1956
Sqn Ldr G. H. S. Tosland, AFC	Jan 1957
Sqn Ldr O. D. Staple, DFC	Sep 1959
Sqn Ldr B. A. Wood, AFC	Dec 1961
Sqn Ldr R. I. F. Garrett	Oct 1963
Sqn Ldr I. A. Hutchins, AFC	Sep 1965
Sqn Ldr W. J. Cranfield, AFC	Nov 1967
Sqn Ldr G. Thompson	Feb 1970
Sqn Ldr D. G. McAllister	Feb 1972
Sqn Ldr R. Davidson	Feb 1974
Wg Cdr J. A. Scrimshaw, AFC	Oct 1975
Wg Cdr G. A. Parkinson	Jan 1976

Disbanded 6 December 1977.

NO. 42 SQUADRON

Formed at Rongotai on 20 December 1943 as No. 42 (Communications) Squadron.

Wg Cdr J. D. Hewett*	Dec 1943
Sqn Ldr K. B. Robinson (acting)	Aug 1945
Wg Cdr J. R. S. Agar	Sep 1945

Disbanded 31 August 1946.

Reformed at Ohakea on 15 March 1950 as No. 42 Squadron.

Sqn Ldr J. R. Wenden	Mar 1950
Sqn Ldr R. D. McVicker, AFC	Mar 1952
Sqn Ldr K. A. Sawyer, DFC	Sep 1955
Sqn Ldr T. J. Danaher, AFC	Nov 1959
Sqn Ldr D. R. H. Thomas	Dec 1962
Sqn Ldr J. C. Evison AFC	Jul 1964
Sqn Ldr B. J. Grigg	Jun 1968
Sqn Ldr A. E. Thomson	Nov 1970
Sqn Ldr F. H. Roach	May 1972
Sqn Ldr G. R. Walsh (acting)	Aug 1974
Sqn Ldr R. T. R. Gilbert, AFC	Sep 1974
Sqn Ldr I. A. Wright	Jun 1977
Sqn Ldr R. L. Randal	Dec 1978
Sqn Ldr D. J. Comber	Dec 1980
Flt Lt R. G. Henderson (acting)	Jan 1983
Sqn Ldr D. M. Hamilton	Mar 1983
Sqn Ldr K. L. Crofskey	Dec 1984
Sqn Ldr A. J. Beattie	Jun 1985

* CO Communications Flight, Rongotai, from June 1940.

NO. 75 SQUADRON

Formed at Goldington (Bedford) in England on 1 October 1916 as No. 75 Squadron, Royal Flying Corps (later Royal Air Force).

Maj H. A. Petre, RFC	Oct 1916
Maj T. O. B. Hubbard, RFC	Jul 1917
Maj T. F. Rutledge, RFC	Jul 1917
Maj C. S. Ross, RFC	Oct 1917
Maj C. A. Ridley, RFC	Nov 1918

Disbanded 13 June 1919.

Reformed at Feltwell (Norfolk) on 15 March 1937 as No. 75 Squadron, RAF.

Sqn Ldr H. L. Rough, DFC, RAF	Mar 1937
Wg Cdr K. E. Ward, RAF	Oct 1937
Wg Cdr E. S. Goodwin, AFC, RAF	Sep 1938
Wg Cdr D. D'A. A. Greig, DFC, AFC, RAF	Nov 1938

Renamed No. 75 (New Zealand) Squadron, RAF, 4 April 1940.

Wg Cdr M. W. Buckley, MBE	Apr 1940
Wg Cdr C. E. Kay, OBE DFC	Nov 1940
Wg Cdr R. Sawrey-Cookson, DSO, DFC, RAF	Sep 1941
Wg Cdr E. G. Olson	Apr 1942
Wg Cdr V. Mitchell, DFC, RAF	Jul 1942
Wg Cdr G. A. Lane, DFC, RAF	Jan 1943
Wg Cdr M. Wyatt, DFC, RAF	May 1943
Wg Cdr R. D. Max, DFC	Aug 1943
Wg Cdr R. J. A. Leslie, AFC	May 1944
Wg Cdr R. J. Newton, DFC	Dec 1944
Wg Cdr C. H. Baigent, DSO, DFC (and bar)	Jan 1945

Disbanded 15 October 1945.

Reformed at Ohakea on 11 October 1946 as No. 75 Squadron.

Wg Cdr J. E. Watts, AFC	Oct 1946
Wg Cdr A. A. N. Breckon, DFC	Apr 1947
Sqn Ldr C. H. Baigent, DSO, DFC (and bar)	Apr 1947
Sqn Ldr R. F. Watson, AFC (acting)	May 1947
Sqn Ldr C. H. Baigent, DSO, DFC (and bar), AFC	Sep 1947
Vacant	Feb 1950*
Sqn Ldr E. C. Gartrell, OBE, DFC	May 1950
Sqn Ldr A. F. Tucker	May 1952
Sqn Ldr N. H. Bright	May 1953
Wg Cdr D. F. St George, DFC	May 1954
Sqn Ldr N. F. Curtis, RAF	Aug 1954
Sqn Ldr G. Amor, RAF	Aug 1956
Sqn Ldr G. R. B. Highet, DFC, AFC	Jul 1958
Sqn Ldr H. G. Moss	Jan 1961
Sqn Ldr I. M. Gillard	Apr 1962
Sqn Ldr B. Stanley-Hunt, AFC	Apr 1962
Sqn Ldr D. M. Dallison	Jun 1965
Sqn Ldr C. W. Rudd	Dec 1965
Sqn Ldr J. A. Scrimshaw, AFC	Dec 1967
Sqn Ldr W. R. Donaldson	Dec 1968
Sqn Ldr J. A. Scrimshaw, AFC	Jun 1969
Sqn Ldr J. S. Boys, AFC	Jan 1970
Sqn Ldr J. A. Scrimshaw, AFC	May 1970
Wg Cdr F. M. Kinvig, AFC	Apr 1971
Wg Cdr J. S. Boys, AFC	May 1973
Wg Cdr R. F. Lawry, AFC	Feb 1975
Wg Cdr C. C. McAllister	Feb 1977
Sqn Ldr G. J. W. Goldsmith (acting)	Mar 1978
Wg Cdr J. S. Hosie	Jun 1978
Wg Cdr G. J. W. Goldsmith, AFC	Dec 1979
Wg Cdr J. W. Lanham	Dec 1981
Wg Cdr J. S. Barclay	Dec 1983
Wg Cdr F. S. Sharp, AFC	Nov 1985

* There was no Commanding Officer appointed during this period.

Appendix B

The 'New Zealand' Squadrons in the Royal Air Force.

NO. 485 SQUADRON
Formed at Driffield on 1 March 1941 as No. 485 (NZ) Squadron.

Sqn Ldr M. W. B. Knight, RAF	Mar 1941
Sqn Ldr E. P. Wells, RAF	Nov 1941
Sqn Ldr R. J. C. Grant, DFC, DFM	May 1942
Sqn Ldr R. W. Baker, DFC	Mar 1943
Sqn Ldr J. M. Checketts, DFC	Jul 1943
Sqn Ldr M. R. D. Hume	Sep 1943
Sqn Ldr J. B. Niven, RAF	Feb 1944
Sqn Ldr J. G. Pattison, DFC	Sep 1944
Sqn Ldr K. J. Macdonald	Feb 1945
Sqn Ldr S. F. Browne, DFC	Jul 1945

Disbanded 26 August 1945.

NO. 486 SQUADRON
Formed at Kirton-in-Lindsey on 7 March 1942 as No. 486 (NZ) Squadron.

Sqn Ldr C. L. C. Roberts, RAF	Mar 1942
Sqn Ldr D. J. Scott, DSO, DFC	Apr 1943
Sqn Ldr I. D. Waddy	Sep 1943
Sqn Ldr J. H. Iremonger, RAF	Jan 1944
Sqn Ldr A. E. Umbers, DFC	Dec 1944
Sqn Ldr K. G. Taylor-Cannon, DFC	Feb 1945
Sqn Ldr W. E. Schrader, DFC	Apr 1945
Sqn Ldr C. J. Sheddan	May 1945

Disbanded 12 October 1945.

NO. 487 SQUADRON
Formed at Feltwell on 15 August 1942 as No. 487 (NZ) Squadron.

Wg Cdr F. C. Seavill, RAF	Aug 1942
Wg Cdr G. J. Grindell, RAF	Dec 1942
Wg Cdr A. G. Wilson, RAF	May 1943
Wg Cdr I. S. Smith, RAF	Feb 1944
Wg Cdr R. C. Porteous, RAF	Aug 1944
Wg Cdr R. W. Baker, DFC	Jan 1945
Wg Cdr F. H. Denton, RAF	Feb 1945
Wg Cdr W. P. Kemp, DFC	Aug 1945

Disbanded 19 September 1945.

NO. 488 SQUADRON
Formation as No. 488 (NZ) Squadron began at Rongotai (Wellington) in September 1941 and completed in Singapore October–November 1941.

Sqn Ldr W. G. Clouston, DFC, RAF	Sep 1941
Sqn Ldr J. N. MacKenzie, DFC, RAF	Jan 1942

Returned to New Zealand after fall of Singapore.
Reformed at Church-Fenton on 25 June 1942 as No. 488 (NZ) Squadron.

Wg Cdr R. M. Trousdale, RAF	Jun 1942
Wg Cdr J. Nesbitt-Dufort, RAF	Mar 1943
Wg Cdr A. R. Burton-Giles, RAF	Jul 1943
Wg Cdr P. H. Hamley, RAF	Sep 1943
Wg Cdr R. C. Haine, RAF	Jan 1944
Wg Cdr R. G. Watts	Oct 1944

Disbanded 26 April 1945.

NO. 489 SQUADRON
Formed at Leuchars on 12 August 1941 as No. 489 (NZ) Squadron.

Wg Cdr J. A. S. Brown, RAF	Aug 1941
Wg Cdr V. C. Darling, RAF	Oct 1942
Wg Cdr J. S. Dinsdale, DSO, DFC	Aug 1943
Wg Cdr L. A. Robertson	Aug 1944
Wg Cdr D. H. Hammond, DSO, DFC	Feb 1945

Disbanded 1 August 1945.

NO. 490 SQUADRON
Formed at Jui on 28 March 1943 as No. 490 (NZ) Squadron.

Wg Cdr D. W. Baird, AFC	Jun 1943
Wg Cdr B. S. Nicholl	Nov 1943
Wg Cdr T. F. Gill, DSO	Oct 1944

Disbanded 1 August 1945.

Appendix C
Chiefs of the Air Staff

Group Captain the Hon R. A. Cochrane, CBE, AFC, RAF
April 1937–March 1939

Air Vice-Marshal I. G. Morrison, CB, CBE
July 1962–July 1966

Air Commodore H. W. L. Saunders, CBE, MC, DFC, MM, RAF
March 1939–December 1941

Air Vice-Marshal C. A. Turner, CB, CBE
July 1966–July 1969

Air Vice-Marshal R. V. Goddard, CB, CBE, RAF
December 1941–July 1943

Air Vice-Marshal W. H. Stratton, CB CBE, DFC
July 1969–July 1971

Air Vice-Marshal Sir Leonard M. Isitt, KBE
July 1943–May 1946

Air Vice-Marshal D. F. St. George, CB, CBE, DFC, AFC
July 1971–September 1974

Air Vice-Marshal Sir Arthur de T. Nevill, KBE, CB
May 1946–January 1951

Air Vice-Marshal R. B. Bolt, CBE, DFC, AFC
September 1974–October 1976

Air Vice-Marshal D. V. Carnegie, CB, CBE, AFC, RAF
January 1951–February 1954

Air Vice-Marshal C. L. Siegert, CB, CBE, MVO, DFC, AFC
October 1976–October 1979

Air Vice-Marshal W. H. Merton, CB, OBE, RAF
February 1954–June 1956

Air Vice-Marshal D. E. Jamieson, CB, OBE
October 1979–April 1983

Air Vice-Marshal C. E. Kay, CB, CBE, DFC
June 1956–July 1958

Air Vice-Marshal D. M. Crooks, CB, OBE
April 1983–October 1986

Air Vice-Marshal M. F. Calder, CB, CBE
July 1958–July 1962

Air Vice-Marshal P. Neville, OBE, AFC
October 1986–

Appendix D

New Zealand Military Aircraft (In order of entry into service)

QTY	TYPE AND VARIANTS DESCRIPTION/MAIN ROLES/REMARKS	PERIOD OF SERVICE	ENGINE POWER/MODEL/DESCRIPTION WINGSPAN/LENGTH/MAXIMUM WEIGHT/MAXIMUM SPEED
1	Bleriot XI-2 general purpose	1913–1914	80 hp Gnome, rotary piston 31 ft 0 in, 28 ft 0 in, 1,760 lbs, 90 mph
7	Bristol F2B Fighter general purpose, army co-op (Imperial gift)	1919–1936	275 hp Rolls Royce Falcon, liquid cooled piston 39 ft 3 in, 25 ft 10 in, 2,800 lbs, 125 mph @ sl
2	de Havilland DH.4 advanced pilot trainer (Imperial gift)	1919–1929	375 hp Rolls Royce Eagle, liquid cooled piston 42 ft 5 in, 30 ft 8 in, 3,472 lbs, 136 mph @ 7,000 ft
27	Avro 504K pilot trainer (Imperial gift except 6)	1920–1930	100 hp Gnome Monosoupape, rotary piston 36 ft 0 in, 29 ft 5 in, 1,800 lbs, 82 mph @ 7,000 ft
9	de Havilland DH.9 advanced pilot trainer (Imperial gift)	1923–1929	230 hp Siddeley Puma, liquid cooled piston 42 ft 5 in, 30 ft 5 in, 3,325 lbs, 111 mph @ 10,000 ft
1	de Havilland DH.50A transport, aerial survey	1927–1931	242 hp Siddeley Puma, liquid cooled piston 42 ft 11 in, 30 ft 5 in, 4,200 lbs, 109 mph
3	Gloster Grebe fighter, advanced pilot trainer	1928–1938	400 hp Armstrong Siddeley Jaguar, radial piston 29 ft 4 in, 20 ft 3 in, 2,614 lbs, 152 mph @ sl
7	de Havilland DH.60 Moth elementary pilot trainer, utility (including floatplane)	1929–1939	100 hp de Havilland Gipsy, in-line piston 30 ft 0 in, 23 ft 11 in, 1,750 lbs, 102 mph
3	Fairey IIIF general purpose, navy co-op floatplane	1929–1939	570 hp Napier Lion, liquid cooled piston 45 ft 9 in, 35 ft 6 in, 6,301 lbs, 120 mph @ 10,000 ft
1	Saro A-17M Cutty Sark utility transport flying boat	1930–1936	2 × 104 hp ADC Cirrus Hermes, in-line piston 45 ft 0 in, 34 ft 11 in, 3,500 lbs, 100 mph
4	Hawker Tomtit elementary and intermediate pilot trainer	1931–1939	150 hp Armstrong Siddeley Mongoose, radial piston 28 ft 7 in, 23 ft 8 in, 1,750 lbs, 124 mph @ 1,000 ft
1	de Havilland DH.80A Puss Moth aerial survey, utility	1931–1936	120 hp de Havilland Gipsy, in-line piston 36 ft 9 in, 25 ft 0 in, 2,050 lbs, 128 mph
39	Vickers Vildebeest III, IV general reconnaissance, advanced pilot trainer, target tug	1935–1944	635 hp Bristol Pegasus, radial piston 49 ft 0 in, 36 ft 8 in, 8,500 lbs, 143 mph
4	Avro 626 advanced pilot and crew trainer	1935–1943	277 hp Armstrong Siddeley Cheetah, radial piston 34 ft 0 in, 26 ft 6 in, 2,667 lbs, 130 mph @ sl
**	Supermarine Walrus I naval reconnaissance amphibian (NZ Division Royal Navy)	1936–1942	775 hp Bristol Pegasus, radial piston 45 ft 10 in, 37 ft 7 in, 7,200 lbs, 135 mph @ 5,000 ft
1	Percival P.1 Gull Four utility	1937–1940	130 hp de Havilland Gipsy Major, in-line piston 36 ft 0 in, 24 ft 6 in, 2,300 lbs, 153 mph
29	Blackburn Baffin general reconnaissance	1937–1941	580 hp Bristol Pegasus, radial piston 45 ft 7 in, 38 ft 4 in, 7,710 lbs, 125 mph @ sl
302	Airspeed AS.10 Oxford I, II multi-engine pilot trainer, crew trainer	1938–1952	2 × 375 hp Armstrong Siddeley Cheetah, radial piston 53 ft 4 in, 34 ft 6 in, 7,800 lbs, 188 mph @ 8,000 ft
5	Vickers Wellington I medium bomber (not delivered to NZ)	1939 only	2 × 1050 hp Bristol Pegasus, radial piston 86 ft 0 in, 61 ft 3 in, 24,850 lbs, 250 mph @ 15,000 ft
41	Fairey Gordon I, II advanced pilot trainer, target tug	1939–1943	525 hp Armstrong Siddeley Panther, radial piston 45 ft 9 in, 36 ft 9 in, 5,906 lbs, 145 mph @ 3,000 ft
62	Vickers Vincent general reconnaissance, advanced pilot trainer, target tug	1939–1944	635 hp Bristol Pegasus, radial piston 49 ft 0 in, 36 ft 8 in, 8,100 lbs, 142 mph @ 5,000 ft
4	Miles M.2F, H, P Hawk Major impressed light monoplane	1939–1943	130 hp de Havilland Gipsy Major, in-line piston 33 ft 0 in, 24 ft 0 in, 1,800 lbs, 150 mph
22	de Havilland DH.60 Moth, Moth Major impressed elementary pilot trainer	1939–1943	100 hp de Havilland Gipsy, in-line piston 30 ft 0 in, 23 ft 11 in, 1,750 lbs, 102 mph
1	Waco UOC impressed cabin biplane	1939–1946	210 hp Continental R-670, radial piston 35 ft 0 in, 25 ft 4 in, 3,100 lbs, 150 mph
3	Miles M.11A Whitney Straight impressed cabin monoplane	1939–1946	130 hp de Havilland Gipsy Major, in-line piston 35 ft 8 in, 25 ft 0 in, 1,896 lbs, 145 mph
2	de Havilland DH.84 Dragon II impressed airliner; radio trainer, transport	1939–1943	2 × 130 hp de Havilland Gipsy Major, in-line piston 47 ft 4 in, 34 ft 6 in, 4,500 lbs, 128 mph @ sl
3	de Havilland DH.86 Express Airliner impressed airliner; radio and navigation trainer, reconnaissance, transport	1939–1945	4 × 200 hp de Havilland Gipsy Six, in-line piston 64 ft 6 in, 46 ft 1 in, 10,150 lbs, 166 mph @ sl
1	Percival P.10 Vega Gull impressed cabin monoplane	1939–1946	205 hp de Havilland Gipsy Six, in-line piston 39 ft 6 in, 25 ft 4 in, 3,250 lbs, 173 mph
1	Waco UIC impressed cabin biplane	1939–1946	210 hp Continental R-670, radial piston 33 ft 3 in, 25 ft 2 in, 2,800 lbs, 140 mph

2	Miles M.14A Hawk Trainer (Magister) impressed elementary trainer	1939–1946	130 hp de Havilland Gipsy Major, in-line piston 33 ft 10 in, 24 ft 8 in, 1,900 lbs, 145 mph @ 1,000 ft
5	de Havilland DH.89, 89A Dragon Rapide impressed airliner; navigation trainer, reconnaissance, transport	1939–1944	2 × 200 hp de Havilland Gipsy Six, in-line piston 48 ft 0 in, 34 ft 6 in, 5,500 lbs, 157 mph @ 1,000 ft
1	BA L.25c Swallow II impressed light monoplane	1939–1940	90 hp Blackburn Cirrus Minor, in-line piston 42 ft 9 in, 27 ft 0 in, 1,500 lbs, 104 mph
4	de Havilland DH.80A Puss Moth impressed cabin monoplane	1939–1946	130 hp de Havilland Gipsy Major, in-line piston 36 ft 9 in, 25 ft 0 in, 2,050 lbs, 128 mph
1	Foster-Wikner Wicko GM.1 impressed cabin monoplane	1939–1942	130 hp de Havilland Gipsy Major, in-line piston 34 ft 6 in, 23 ft 3 in, 2,000 lbs, 140 mph
335	de Havilland DH.82A Tiger Moth II elementary pilot trainer (some impressed)	1939–1956	130 hp de Havilland Gipsy Major, in-line piston 29 ft 4 in, 23 ft 11 in, 1,825 lbs, 109 mph @ sl
1	General Aircraft Monospar ST-25 Jubilee impressed cabin monoplane	1939–1941	2 × 90 hp Pobjoy Niagara, radial piston 40 ft 2 in, 26 ft 4 in, 2,875 lbs, 142 mph @ sl
1	Beech C.17L impressed cabin biplane	1939–1946	225 hp Jacobs L-4, radial piston 32 ft 0 in, 24 ft 5 in, 3,150 lbs, 175 mph @ sl
5	de Havilland DH.94 Moth Minor impressed light monoplane	1940–1946	90 hp de Havilland Gipsy Minor, in-line piston 36 ft 7 in, 24 ft 5 in, 1,550 lbs, 118 mph
1	Waco QDC impressed cabin biplane	1940–1941	165 hp Continental A-70, radial piston 33 ft 3 in, 23 ft 2 in, 2,581 lbs, 116 mph
1	Porterfield 35-W impressed cabin monoplane	1940 only	90 hp Warner Scarab Junior, radial piston 32 ft 0 in, 20 ft 1 in, 1,326 lbs, 133 mph
63	Hawker Hind, Hind Trainer army co-op, advanced pilot trainer	1940–1943	640 hp Rolls Royce Kestrel, liquid cooled piston 37 ft 3 in, 29 ft 7 in, 5,298 lbs, 186 mph @ 16,000 ft
202	North American Harvard II, IIA, IIB, III advanced pilot trainer	1941–1977	550 hp Pratt & Whitney R-1340 Wasp, radial piston 42 ft 0 in, 29 ft 0 in, 5,600 lbs, 205 mph @ 5,000 ft
94	Lockheed Hudson III, IIIA, V, VI general reconnaissance bomber	1941–1948	2 × 1200 hp Wright R-1820 Cyclone Nine, radial piston 65 ft 6 in, 44 ft 4 in, 20,500 lbs, 253 mph @ 15,000 ft
**	Brewster 339E Buffalo I fighter; defence of Singapore	1941–1942	1100 hp Wright R-1820 Cyclone Nine, radial piston 35 ft 0 in, 26 ft 0 in, 4,479 lbs, 292 mph @ 20,000 ft
4	Short Singapore III maritime reconnaissance flying boat	1941–1943	4 × 630 hp Rolls Royce Kestrel, liquid cooled piston 90 ft 0 in, 64 ft 2 in, 31,500 lbs, 145 mph @ 2,000 ft
**	Hawker Hurricane IIB fighter; defence of Singapore	1942 only	1280 hp Rolls Royce Merlin, liquid cooled piston 40 ft 0 in, 32 ft 0 in, 8,250 lbs, 340 mph @ 22,000 ft
297	Curtiss P-40E, K, L, M, N Kittyhawk/Warhawk fighter, fighter bomber	1942–1945	1200 hp Allison V-1710, liquid cooled piston 37 ft 4 in, 33 ft 4 in, 8,850 lbs, 378 mph @ 11,000 ft
23	Avro 652A Anson I navigation trainer, radio research	1942–1952	2 × 340 hp Armstrong Siddeley Cheetah, radial piston 56 ft 6 in, 42 ft 3 in, 8,000 lbs, 188 mph @ 7,000 ft
4	Rearwin 9000L, KR Sportster impressed cabin monoplane	1942–1946	90 hp Ken Royce 5F or Le Blond 5F, radial piston 35 ft 0 in, 22 ft 3 in, 1,460 lbs, 120 mph @ 1,000 ft
1	Porterfield 35-W re-impressed cabin monoplane	1942–1946	90 hp Warner Scarab Junior, radial piston 32 ft 0 in, 20 ft 1 in, 1,326 lbs, 133 mph
11	Supermarine Walrus I maritime pilot trainer amphibian	1943–1946	775 hp Bristol Pegasus, radial piston 45 ft 10 in, 37 ft 7 in, 7,200 lbs, 135 mph @ 5,000 ft
49	Douglas C-47, -47A, -47B, Dakota medium range/VIP transport, target tug	1943–1977	2 × 1200 hp Pratt & Whitney R-1830 Twin Wasp, radial piston 95 ft 0 in, 64 ft 6 in, 31,000 lbs, 230 mph @ 9,000 ft
56	Consolidated PBY-5, PB2B-1 Catalina maritime reconnaissance, air sea rescue flying boat	1943–1954	2 × 1200 hp Pratt & Whitney R-1830 Twin Wasp, radial piston 104 ft 0 in, 63 ft 10 in, 34,000 lbs, 190 mph @ 11,000 ft
1	de Havilland DH.83 Fox Moth impressed cabin biplane	1943-1948	130 hp de Havilland Gipsy Major, in-line piston 30 ft 11 in, 25 ft 9 in, 2,070 lbs, 123 mph
9	Lockheed C-60A Lodestar medium range transport	1943–1948	2 × 1200 hp Wright R-1820 Cyclone Nine, radial piston 65 ft 6 in, 49 ft 10 in, 21,000 lbs, 266 mph @ 17,000 ft
139	Lockheed PV-1, B-34A Ventura general reconnaissance bomber	1943–1946	2 × 2000 hp Pratt & Whitney R-2800 Double Wasp, radial piston 65 ft 6 in, 51 ft 9 in, 34,000 lbs, 322 mph @ 14,000 ft
69	Douglas SBD-3, -4, -5 Dauntless dive bomber	1943–1944	1200 hp Wright R-1820 Cyclone Nine, radial piston 41 ft 6 in, 33 ft 0 in, 10,855 lbs, 255 mph @ 14,000 ft
9	de Havilland DH.89B Dominie II light transport	1943–1953	2 × 200 hp de Havilland Gipsy Queen, in-line piston 48 ft 0 in, 34 ft 6 in, 5,500 lbs, 157 mph @ 1,000 ft
48	Grumman TBF-1, -1C Avenger light bomber, target tug, topdressing trials	1943–1959	1700 hp Wright R-2600 Cyclone Fourteen, radial piston 54 ft 2 in, 40 ft 0 in, 16,000 lbs, 271 mph @ 12,000 ft
424	Chance Vought F4U-1, -1D, FG-1D Corsair fighter, fighter bomber	1944–1948	2250 hp Pratt & Whitney R-2800 Double Wasp, radial piston 41 ft 0 in, 33 ft 5 in, 14,000 lbs, 417 mph @ 20,000 ft
4	Short Sunderland III long range transport flying boat	1944–1947	4 × 1065 hp Bristol Pegasus, radial piston 112 ft 10 in, 85 ft 4 in, 58,000 lbs, 210 mph @ 7,000 ft

4	Lockheed PV-2 Harpoon general reconnaissance bomber	1945 only	2 × 2000 hp Pratt & Whitney R-2800 Double Wasp, radial piston 75 ft 0 in, 52 ft 1 in, 36,000 lbs, 282 mph @ 14,000 ft
1	Mitsubishi Reisen 22 (Zeke) surrendered enemy fighter	1945 only	1130 hp Nakajima NK1F Sakae, radial piston 39 ft 5 in, 29 ft 9 in, 5,906 lbs, 332 mph @ 17,000 ft
1	Gloster Meteor III fighter	1945–1950	2 × 2000 lbs thrust Rolls Royce Derwent, turbojet 43 ft 0 in, 41 ft 3 in, 14,750 lbs, 475 mph @ 30,000 ft
2	Avro 652A Anson XII VIP transport (for UK High Commission)	1946–1950	2 × 420 hp Armstrong Siddeley Cheetah, radial piston 56 ft 6 in, 42 ft 3 in, 10,500 lbs, 190 mph
85	de Havilland DH.98 Mosquito T.III, FB.VI, FB.40, T.43 fighter bomber	1946–1952	2 × 1640 hp Rolls Royce Merlin, liquid cooled piston 54 ft 2 in, 40 ft 10 in, 22,000 lbs, 378 mph @ 13,000 ft
6	Auster J/5 army observation, light utility	1947–1969	130 hp de Havilland Gipsy Major, in-line piston 36 ft 0 in, 23 ft 5 in, 2,150 lbs, 114 mph
30	de Havilland DH.104 Devon light transport, pilot, navigation and radio trainer	1948–1981	2 × 355 hp de Havilland Gipsy Queen, in-line piston 57 ft 0 in, 39 ft 3 in, 8,800 lbs, 220 mph @ sl
6	Airspeed AS.65 Consul light transport (converted from Oxfords)	1948–1952	2 × 375 hp Armstrong Siddeley Cheetah, radial piston 53 ft 4 in, 35 ft 6 in, 8,250 lbs, 190 mph @ 5,000 ft
2	Miles M.57 Aerovan IV aerial topdressing trials, magnetic survey	1949–1951	2 × 150 hp Blackburn Cirrus Major, in-line piston 50 ft 0 in, 36 ft 0 in, 5,800 lbs, 127 mph @ sl
47	de Havilland DH.100 Vampire FB.5, FB.52 fighter bomber	1951–1973	3100 lbs thrust de Havilland Goblin, turbojet 38 ft 0 in, 30 ft 9 in, 12,360 lbs, 531 mph @ sl
30	North American P-51D Mustang fighter bomber (in storage from 1945)	1951–1957	1650 hp Packard V-1650 Merlin, liquid cooled piston 37 ft 0 in, 32 ft 3 in, 11,600 lbs, 437 mph @ 25,000 ft
12	Bristol 170 Freighter Mk 31(NZ)M medium range/tactical transport	1951–1977	2 × 2040 hp Bristol Hercules, radial piston 108 ft 0 in, 68 ft 4 in, 46,000 lbs, 230 mph @ 3,000 ft
4	Handley Page HP.95 Hastings C.3 long range transport	1952–1965	4 × 2090 hp Bristol Hercules, radial piston 112 ft 9 in, 82 ft 1 in, 83,000 lbs, 350 mph
11	de Havilland DH.115 Vampire T.11, T.55 advanced pilot trainer	1952–1973	3500 lbs thrust de Havilland Goblin, turbojet 38 ft 0 in, 34 ft 7 in, 11,150 lbs, 549 mph @ 20,000 ft
**	de Havilland DH.100 Vampire FB.9 fighter bomber; Cyprus, Singapore	1952–1956	3350 lbs thrust de Havilland Goblin, turbojet 38 ft 0 in, 30 ft 9 in, 12,390 lbs, 548 mph @ 30,000 ft
**	Gloster Meteor T.7 instrument continuation trainer; Cyprus	1952–1955	2 × 3500 lbs thrust Rolls Royce Derwent, turbojet 37 ft 2 in, 43 ft, 6 in, 17,600 lbs, 585 mph @ sl
16	Short Sunderland MR.5 maritime reconnaissance flying boat	1953–1967	4 × 1200 hp Pratt & Whitney R-1830 Twin Wasp, radial piston 112 ft 10 in, 85 ft 4 in, 65,000 lbs, 213 mph @ 5,000 ft
**	de Havilland DH.115 Vampire T.11 instrument continuation trainer; Cyprus, Singapore	1953–1958	3500 lbs thrust de Havilland Goblin, turbojet 38 ft 0 in, 34 ft 7 in, 11,150 lbs, 549 mph @ 20,000 ft
**	de Havilland DH.112 Venom FB.1 fighter bomber; Singapore	1955–1958	4850 lbs thrust de Havilland Ghost, turbojet 41 ft 8 in, 31 ft 10 in, 15,400 lbs, 640 mph
1	Auster Mk 7c Antarctic Antarctic light utility, army observation	1956–1966	145 hp de Havilland Gipsy Major, in-line piston 36 ft 0 in, 23 ft 9 in, 2,122 lbs, 124 mph @ 1,000 ft
1	de Havilland DHC.2 Beaver Antarctic light transport	1956–1960	450 hp Pratt & Whitney R-985 Wasp Junior, radial piston 48 ft 0 in, 30 ft 5 in, 5,100 lbs, 163 mph @ 5,000 ft
**	English Electric Canberra B.2, T.4 medium bomber; Singapore	1958–1962	2 × 6500 lbs thrust Rolls Royce Avon, turbojet 64 ft 0 in, 65 ft 6 in, 46,000 lbs, 570 mph @ 40,000 ft
11	English Electric Canberra B(I).12 strike bomber	1959–1970	2 × 7400 lbs thrust Rolls Royce Avon, turbojet 64 ft 0 in, 65 ft 6 in, 54,950 lbs, 541 mph @ 40,000 ft
1	de Havilland DHC.3 Otter Antarctic light transport (mainly in storage)	1960–1963	600 hp Pratt & Whitney R-1340 Wasp, radial piston 58 ft 0 in, 41 ft 10 in, 8,000 lbs, 160 mph @ 5,000 ft
2	English Electric Canberra T.13 type conversion trainer	1961–1970	2 × 6500 lbs thrust Rolls Royce Avon, turbojet 64 ft 0 in, 65 ft 6 in, 46,000 lbs, 518 mph
3	Douglas DC-6 long range transport	1961–1968	4 × 2400 hp Pratt & Whitney R-2800 Double Wasp, radial piston 117 ft 6 in, 100 ft 7 in, 97,200 lbs, 356 mph @ 20,000 ft
5	Lockheed C-130H Hercules long range/tactical transport	1965–	4 × 4508 eshp Allison T-56, turboprop 132 ft 7 in, 99 ft 6 in, 175,000 lbs, 386 mph @ 25,000 ft
13	Bell 47G-3B1, -3B2 Sioux army observation, pilot trainer helicopter	1965–	280 hp Lycoming TVO-435, horizontally opposed piston 37 ft 2 in, 43 ft 3 in, 2,950 lbs, 105 mph @ sl
15	Bell UH-1D, H Iroquois utility transport helicopter	1966–	1400 shp Lycoming T-53, free turbine 48 ft 3 in, 57 ft 1 in, 9,500 lbs, 127 mph
6	Lockheed P-3B Orion maritime reconnaissance	1966–	4 × 4910 eshp Allison T-56, turboprop 99 ft 8 in, 116 ft 10 in, 139,760 lbs, 476 mph @ 15,000 ft
8	Westland Wasp HAS.1 anti-submarine helicopter	1966–	710 shp Bristol Siddeley Nimbus, free turbine 32 ft 3 in, 40 ft 4 in, 5,500 lbs, 120 mph @ sl
4	AESL Airtourer T6/24 elementary pilot trainer	1970–	150 hp Lycoming O-320, horizontally opposed piston 26 ft 0 in, 21 ft 6 in, 1,900 lbs, 164 mph @ sl

18 McDonnell Douglas A-4G, K Skyhawk fighter ground attack, maritime strike	1970–	9300 lbs thrust Pratt & Whitney J-52, turbojet 27 ft 6 in, 41 ft 3 in, 24,500 lbs, 674 mph @ sl
6 McDonnell Douglas TA-4G, K Skyhawk type conversion trainer, fighter ground attack, maritime strike	1970–	9300 lbs thrust Pratt & Whitney J-52, turbojet 27 ft 6 in, 43 ft 7 in, 24,500 lbs, 674 mph @ sl
16 BAC Strikemaster Mk 88 advanced pilot trainer	1972–	3410 lbs thrust Bristol Siddeley Viper, turbojet 36 ft 11 in, 33 ft 9 in, 11,500 lbs, 450 mph @ sl
19 NZ Aerospace CT-4B Airtrainer basic pilot trainer	1976–	210 hp Teledyne Continental IO-360, horizontally opposed piston 26 ft 0 in, 23 ft 2 in, 2,650 lbs, 178 mph @ sl
10 Hawker Siddeley HS.748MF Andover C.1 medium range/tactical/VIP transport	1976–	2 × 3245 eshp Rolls Royce Dart, turboprop 98 ft 3 in, 78 ft 0 in, 55,000 lbs, 320 mph @ 15,000 ft
3 Fokker F.27 Friendship Mk 120 navigation and radio trainer, maritime reconnaissance	1980–	2 × 1860 eshp Rolls Royce Dart, turboprop 95 ft 2 in, 77 ft 4 in, 40,500 lbs, 280 mph
3 Cessna 421C Golden Eagle light VIP transport	1981–	2 × 375 hp Continental GTSIO-520, horizontally opposed piston 41 ft 2 in, 36 ft 5 in, 7,450 lbs, 297 mph @ 20,000 ft
3 Boeing 727-22QC long range transport	1981–	3 × 14000 lbs thrust Pratt & Whitney JT8D, turbofan 108 ft 0 in, 133 ft 2 in, 177,000 lbs, 630 mph @ 22,000 ft
** Bell UH-1H Iroquois utility transport helicopter; Sinai	1982–1986	1400 shp Lycoming T-53, free turbine 48 ft 3 in, 57 ft 1 in, 9,500 lbs, 127 mph

NOTES:
QTY — number procured/delivered. ** indicates aircraft hired from UK forces (or US in case of Sinai).
ROLE — aircraft impressed in wartime, unless otherwise indicated, were used for light transport or utility duties.
PERIOD OF SERVICE — from date first example taken over by RNZAF (or its predecessors) until date type withdrawn from flying service. Some types remained on strength, but in storage, for a few years longer before being put up for disposal. Although only years are given here, the order of entries is based on actual dates of entry into service.
WINGSPAN — main rotor diameter for helicopters. MAXIMUM WEIGHT — maximum permissible. MAXIMUM SPEED — in level flight at altitude quoted. (All engine, dimension, and performance data are given only for the most numerous or significant variant of the aircraft in New Zealand service. Vildebeest IV powered by 825 hp Bristol Perseus; Hudson V and VI powered by two 1200 hp Pratt & Whitney R-1830 Twin Wasp.)
ABBREVIATIONS: hp = horsepower, shp = shaft horsepower, eshp = equivalent shaft horsepower, lbs = pounds, ft = feet, in = inches, mph = statute miles per hour, sl = sea level

Bibliography

Bentley, Geoffrey. *RNZAF: A Short History*, A.H. & A.W. Reed, Wellington, 1969.

Bentley, Geoffrey (ed.). *RNZAF News*, December 1958–May 1962.

Darby, Charles. *RNZAF, The First Decade, 1937–46*, Kookaburra Technical Publications Pty Ltd, Melbourne, 1978.

Dean, H. R. *The RNZAF in South-East Asia, 1941–42*, War History Branch, Internal Affairs Department, Wellington, 1952.

Harleyford Publications Ltd. *Lancaster, the Story of a Famous Bomber*, England, 1964.

Harvie, E. F. *George Bolt, Pioneer Aviator*, A.H. & A.W. Reed, Wellington, 1974.

Hunt, Flight Lieutenant Leslie. *Defence Until Dawn*, Washburn & Sons Ltd, Southend-on-Sea, 1949.

Jenks, C. F. L. (compiler). *New Zealand Military Aircraft and Serial Numbers*, Aviation Historical Society of New Zealand, Wellington, 1980.

Jones, Ira. *Tiger Squadron*, W.H. Allen & Co., London, 1954.

Journals of the Aviation Historical Society of New Zealand.

Mitchell, Alan W. *New Zealanders in the Air War*, George G. Harrap & Co. Ltd, London, 1945.

Moyes, Philip J. R. *Bomber Squadrons of the RAF and their Aircraft*, Macdonald & Co. Ltd, London, 1964.

Mulgan, David. *The Kiwi's First Wings*, Wingfield Press, Wellington, 1960.

New Zealand Wings magazine, 1965–85.

O'Shea, Phillip P. *Honours, Titles, Styles and Precedence in New Zealand*, Government Printer, Wellington, 1977.

Raleigh, Sir Walter. 'The War in the Air', Vol. 1, *Official History of the First World War*, Oxford University Press, 1922.

Ross, Squadron Leader J. M. S. 'Royal New Zealand Air Force', *Official History of the RNZAF in the Second World War*, War History Branch, Internal Affairs Department, Wellington, 1955.

Scott, Group Captain Desmond. *Typhoon Pilot*, Leo Cooper in association with Secker & Warburg, London, 1982.

Sutton, Brigadier D. J., OBE. *The Story of the RASC and RCT, 1945–82*, Leo Cooper in association with Secker & Warburg, London, 1983.

Thompson, Wing Commander H. L. 'New Zealanders with the Royal Air Force', *Official History of New Zealand in the Second World War* (3 vols), War History Branch, Internal Affairs Department, Wellington, 1953, 1956, 1959.

Wells, Kevin W. *An Illustrated History of the New Zealand Spitfire Squadron*, Hutchinson, Auckland, 1984.

White, Leo. *Wingspread*, Unity Press Ltd, Auckland, 1945.

Glossary

AA	anti-aircraft
AFC	Air Force Cross
AOC	Air Officer Commanding
AOC-in-C	Air Officer Commanding-in-Chief
ARVN	Army of the Republic of Vietnam
ATC	Air Training Corps
Avionics	Aviation electronics
BR	Bomber Reconnaissance
CAS	Chief of the Air Staff
CFI	chief flying instructor
CFS	Central Flying School
CO	commanding officer
COMAIRNORSOLS	Commander, Air, Northern Solomons
COMAIRSOPAC	Commander, Air, South Pacific
COMZEAIRTAF	Commander, New Zealand Air Task Force
DSO	Distinguished Service Order
DFC	Distinguished Flying Cross
DFM	Distinguished Flying Medal
EATS	Empire Air Training Scheme
EFTS	Elementary Flying Training School
FAC	forward air controller
FEAF	Far East Air Force
FTS	Flying Training School
GOC	General Officer Commanding
GR	General Reconnaissance
HF	high frequency
MC	Military Cross
MFO	Multi-national Force and Observers
MM	Military Medal
NCO	non-commissioned officer
NZAF	New Zealand Air Force
NZCF	New Zealand Cadet Forces
NZPAF	New Zealand Permanent Air Force
OC	officer commanding
OTU	operational training unit
RAAF	Royal Australian Air Force
RAF	Royal Air Force
RCAF	Royal Canadian Air Force
RFC	Royal Flying Corps
RNAS	Royal Naval Air Service
RNZAF	Royal New Zealand Air Force
RNZAF SUS	Royal New Zealand Air Force Support Unit Singapore
RNZIR	Royal New Zealand Infantry Regiment
R/T	radio telephony
RWAU	Rotary Wing Aviation Unit
SEATO	South East Asia Treaty Organisation
SFTS	Service Flying Training School
SSC	Short Service Commission
STOL	short take-off and landing
SU	servicing unit
TAF	Territorial Air Force
TTS	Technical Training School
UHF	ultra-high frequency
USAAF	United States Army Air Force
USAF	United States Air Force
USMC	United States Marine Corps
USN	United States Navy
VC	Victoria Cross
VCAS	Vice-Chief of the Air Staff
VHF	very high frequency
WAAF	Women's Auxiliary Air Force
WRNZAF	Women's Royal New Zealand Air Force
W/T	wireless telegraphy

Index

Page numbers in italics refer to illustrations

SQUADRONS

AIRCRAFT

ROYAL NEW ZEALAND AIR FORCE

PER ARDUA AD ASTRA